AGE OF PROGRESS

TIME
LIFE
BOOKS
®

GREAT AGES OF MAN

A History of the World's Cultures

AGE OF PROGRESS

by

S. C. BURCHELL

and

The Editors of TIME-LIFE BOOKS

TIME-LIFE BOOKS, NEW YORK

THE AUTHOR: Samuel C. Burchell, formerly an instructor in English at Yale and a specialist in Victorian England, is the editor of the Yale Shakespeare edition of *As You Like It*, published in 1954, and author of *Building the Suez Canal*, published in 1966. A Phi Beta Kappa at Yale, Mr. Burchell studied at Cambridge in England and McGill in Canada. He is a frequent contributor to national publications, including *South Atlantic Quarterly*, *The Yale Review* and *Town and Country*.

THE CONSULTING EDITOR: Leonard Krieger, now Professor of History at Columbia University, was formerly Professor of History at Yale; Dr. Krieger is the author of *The German Idea of Freedom* and *The Politics of Discretion* and coauthor of *History*, written in collaboration with John Higham and Felix Gilbert.

THE COVER: Highlights of the Age of Progress—the birth of the airplane and automobile, a spidery new architecture, political upheaval, a glittering social life.

TIME-LIFE BOOKS

EDITOR
Maitland A. Edey

EXECUTIVE EDITOR
Jerry Korn

TEXT DIRECTOR   ART DIRECTOR
Martin Mann   Sheldon Cotler

CHIEF OF RESEARCH
Beatrice T. Dobie

PICTURE EDITOR
Robert G. Mason

Assistant Text Directors:
Ogden Tanner, Diana Hirsh
Assistant Art Director: Arnold C. Holeywell
Assistant Chief of Research: Martha T. Goolrick
Assistant Picture Editor: Melvin L. Scott

PUBLISHER
Walter C. Rohrer
General Manager: John D. McSweeney
Business Manager: John Steven Maxwell
Production Manager: Louis Bronzo

Sales Director: Joan D. Manley
Promotion Director: Beatrice K. Tolleris
Managing Director, International: John A. Millington

GREAT AGES OF MAN

SERIES EDITOR: Russell Bourne
Editorial Staff for *Age of Progress:*
Assistant Editor: Carlotta Kerwin
Text Editors: Anne Horan, Betsy Frankel,
William Longgood, Ogden Tanner,
Robert Tschirky
Picture Editor: Isabelle Rubin
Designer: Norman Snyder
Assistant Designer: Ladislav Svatos
Staff Writers: Sam Halper, Frank Kendig,
Marianna Pinchot, John Stanton,
Jeffrey Tarter, Edmund White
Chief Researcher: Peggy Bushong
Researchers: Terry Drucker,
Kathleen Brandes, John Hochmann,
Patricia Huntington, Carol Isenberg,
Paula Norworth, Theo Pascal, Peter Yerkes,
Johanna Zacharias, Arlene Zuckerman

EDITORIAL PRODUCTION
Color Director: Robert L. Young
Assistant: James J. Cox
Copy Staff: Rosalind Stubenberg,
Barbara Hults, Florence Keith
Picture Department: Dolores A. Littles,
Joan Lynch
Traffic: Arthur A. Goldberger
Art Assistants: Anne Landry,
Robert Pellegrini, Mervyn Clay

Valuable aid in preparing this book was given by the following individuals and departments of Time Inc.: Editorial Production, Robert W. Boyd Jr., Margaret T. Fischer; Editorial Reference, Peter Draz; Picture Collection, Doris O'Neil; Photographic Laboratory, George Karas; TIME-LIFE News Service, Murray J. Gart; Correspondents Maria Vincenza Aloisi (Paris), Barbara Moir (London), Ann Natanson (Rome), Elisabeth Kraemer (Bonn) and Traudl Lessing (Vienna).

CONTENTS

INTRODUCTION

Appropriately, Samuel Burchell begins this book with a description of the opening of the Great Exhibition in London in 1851—that most splendid revelation of the 19th Century's belief in the idea of Progress. Prince Albert, the Exhibition's chief sponsor, saw it as evidence of human history's inevitable advance toward "the realization of the unity of mankind." To Albert, the immense technical achievements revealed in the new machinery and other displays, and the great esthetic achievement of the Crystal Palace in which they were housed, were only outward signs of an inward grace. The Exhibition was the symbol of an ethical progress which the whole world was making, and would continue to make.

This simple belief in progress was shaken by the catastrophe of the First World War, and never really recovered from it. No one would ever again look upon the triumphs of science, technology and learning with "fine, careless rapture." Yet even as late as 1920, when John Bagnell Bury published his pioneering book, *The Idea of Progress*, many people still believed sufficiently in the dream to be shocked by Bury's contention that progress was not an inevitable law of nature, like the law of gravitation, but simply an idea made by man, and a relatively new idea at that.

This continuing faith in an inevitable and beneficent Progress was finally destroyed by the disillusionment that followed the First and Second World Wars, a disillusionment caused as much by the wars' mere occurrence as by the horrors of Auschwitz and Belsen, Hiroshima and Nagasaki. Our danger now is the reverse of the 19th Century's belief in the idea of Progress. We are disposed to ignore the era's very real accomplishments.

As a result of advances made in the 19th and early 20th Centuries, people did live longer, fewer children died as infants, and many were better fed, better housed and better educated. The physical unity of the world was made possible by the steamboat, the locomotive, the automobile and the airplane. The unity of science was exemplified by the adaptation, within a few years of its discovery, of Louis Pasteur's work with bacteria in Paris to Joseph Lister's practice of antiseptic surgery in Scotland. Areas of the world previously uninhabitable, or habitable only at a very low level of existence, became easier to live in.

For more than a century the whole world flattered the West by imitating its belief that progress in politics, science and art was a continuous web; that progress was change, and change was always for the better. We know now that it can often be for the worse. But we must not forget that many changes are good ones. It was the pressure for change, for instance, that after 1776 made the "pursuit of happiness" far easier for far more people than it had ever been before. Prince Albert, like other prophets of progress, was too optimistic, but he was not wrong—either in his aims or in his hopes for their achievement.

SIR DENIS BROGAN
Professor of Political Science
Cambridge University

1

GREAT
EXPECTATIONS

As the sun rose on the first day of May, 1851, its early rays revealed a splendid sight. The great city of London was already stirring; streets normally deserted until much later in the day were today rapidly filling up with frock-coated brokers and bankers, ladies of fashion in sweeping skirts, barmaids, laborers, sailors and policemen, all of them on the move.

Englishmen of all classes were hurrying to the opening of the Great Exhibition in Hyde Park. Many were on foot, but some rode in phaetons and victorias, and others found room on crowded omnibuses. The sounds of horses' hooves and iron wheels rang through the graceful squares of Belgravia and the narrow lanes of Clerkenwell alike.

The hubbub was greatest in the vicinity of Hyde Park. There, on a 26-acre expanse of green lawns shaded by English elms, stood the object of the day's excitement—the Crystal Palace, the vast and glittering building that housed the Great Exhibition. Flags from distant countries fluttered from its iron columns, and its glass walls gleamed with intermittent sunlight in the spring morning.

As morning advanced, excitement rose, and with the arrival of Queen Victoria at noon it reached a crescendo. The Queen was dressed for the occasion in pink and silver and adorned with diamonds. Escorted by a troop of the Household Cavalry, she had ridden from Buckingham Palace in a closed carriage with her husband, Prince Albert, and two of their children. To a flourish of trumpets the Royal Family entered the hall, where they were ceremoniously received by the Royal Commissioners of the Exhibition and the Archbishop of Canterbury. Before a vast assemblage of British and foreign dignitaries, Prince Albert read to the Queen a report on the fair, the Archbishop offered a prayer, and one thousand voices sang Handel's *Hallelujah Chorus*. The Queen and her entourage then traversed the length of the hall, followed by statesmen in order of rank. As they walked, the cheers of the crowd drowned out the strains of the organ music that accompanied the procession. Writing later in her journal, Queen Victoria remarked that the event had presented a spectacle "quite like the Coronation, and for *me*, the same anxiety."

Londoners had been looking forward to this day

A GATEWAY TO PROGRESS, *one of the entrances to the Crystal Palace is guarded by two Yeomen of the Guard who await the arrival of Queen Victoria to open the Great Exhibition. Behind the gates is the Crystal Fountain.*

for almost a year, watching the Crystal Palace as it took shape section by section. Thousands had subscribed a total of £75,000 to the success of the undertaking; the Queen herself gave £1,000 and Prince Albert £500; business firms made substantial contributions, and nameless citizens donated half-crowns and shillings. The railroads offered special group excursion rates from outlying districts to London, clubs formed all over the country so that people could take advantage of the low fares, and laborers saved their earnings for the great event. To hold down the crowds on opening day, admission was limited to holders of season tickets, but more than 25,000 of these were sold.

The inspiration for the Exhibition, which was designed to display "the Works of Industry of All Nations," had come largely from Prince Albert, and its theme was "Progress." Progress was a thought much in the minds of 19th Century men—and not without reason. Never had men been more conscious of triumphing over the world about them, and never had they advanced at such a speed. Scarcely half a century before, three quarters of the population of Europe had lived in a rural world of farms, dirt roads and barge canals. Most wealth had come from the soil. Textiles and tools were mainly hand wrought by small craftsmen, as they had been since time out of mind; power came chiefly from wind and water if not from human or animal muscle, and transportation was mainly by horse or sail. There were large cities like London and Paris, Berlin and Vienna, but such future industrial complexes as Liverpool, Manchester and Lille still had populations under 100,000, and Essen, later to be an industrial giant, was a modest town of 3,000.

In little more than 50 years the increasing substitution of machinery for human effort, of inanimate for animate supplies of power, had wrought the Industrial Revolution. The process had been developing slowly for more than three centuries, but toward the close of the 18th Century it had suddenly accelerated. Now, in the middle of the 19th Century, men found themselves living in an exciting world unimagined by their grandfathers. It was a world of iron and coal and steam, of machinery and engines, of railroads, steamships and telegraph wires. Faster travel and better communication were making the earth a smaller place, and ingenious new devices were making life easier—devices like the indoor water closet and the fixed bathtub, the gas cooking range and the refrigerator.

Not only in industry, but in economics, politics and science as well, 19th Century men perceived that a new order was coming into existence. In the cities, the new middle-class capitalists were miraculously accumulating more and more wealth, and with tireless energy they were drawing all of society into the vortex of their enterprise. Workers were earning higher wages, and Her Majesty's Government was beginning to pass legislation that improved working conditions; there was even talk of granting workers the vote. Science was daily adding to man's knowledge of the world he lived in, and with increasing knowledge medicine was on the threshold of great advances in staying death and easing sickness. The marvels of human ingenuity seemed to be limitless.

"The Exhibition of 1851," Prince Albert told the Royal Commissioners, "is to give us a true test and a living picture of the point of development at which the whole of mankind has arrived . . . and a new starting point from which all nations will be able to direct their further exertions."

The first aspect of the Great Exhibition to strike the visitor on that May morning in 1851 was the extraordinary building that housed it. The Crystal Palace enclosed 19 acres of Hyde Park, and including its mezzanine it had almost a million square feet of floor space. It was made of nearly 300,000 panes of glass set in more than 5,000 columns and

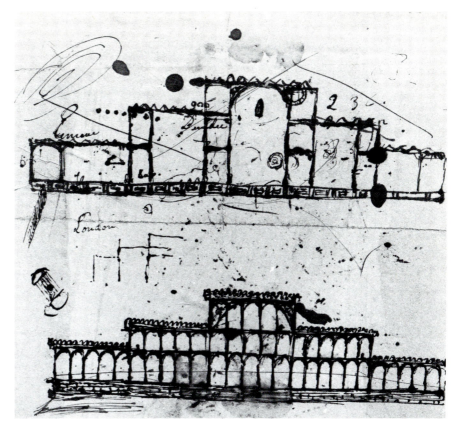

PROPHETIC DOODLES, *the first sketches of what became the Crystal Palace adorn the desk blotter of Joseph Paxton. Although he was not trained as an architect, his idea of a building of glass was accepted after 233 plans submitted by professionals had been rejected. He worked up the final plan in only nine days. The structure's success so delighted the country and Queen Victoria that in the year of the fair he was knighted.*

girders of iron. The glass panes—the largest ever made—were delivered to the site of the Exhibition cut to size; the columns and girders were preshaped in the foundries of Birmingham and delivered by railroad—itself still an exciting novelty—and set into place as little as 18 hours after being cast. This was both one of the first instances of prefabrication and one of the first demonstrations of the possibilities of mass production. It foreshadowed a new age of building.

Architects and engineers all over the British Isles had competed to design the building, and the honor had gone to Joseph Paxton, who at the last minute asked the building committee if he might submit a "notion" he had. Paxton, a former gardener for the Duke of Devonshire, had risen to become a director of the Midland Railway and a notable public figure. While presiding at a railway meeting, he scribbled two sketches on blotting paper, then worked night and day to elaborate on these until he finished the plans for what *Punch* was to dub the "Crystal Palace."

The design was extraordinary for an exhibition hall; it was essentially an outsized version of a greenhouse Paxton had recently designed for the Duke of Devonshire. Could such a structure accommodate thousands of tons of pounding machinery? Would it withstand the weather and the tramp of thousands of feet? Was it even suitable for a great exhibition hall? While the committee puzzled over these questions, Paxton had another idea; he gave his design to the *Illustrated London News*, and thus became the first man to utilize the new picture press to rally public support. He was rewarded with immediate and enthusiastic popular response, and the committee yielded to the public verdict.

One objection remained, and that was to felling the stately elms of Hyde Park. Paxton met this problem by altering his original design to add a great vaulted transept that would roof the elms. Engineers resolved the complex problems of stress posed by the structure, and thousands of laborers were hired to erect it. When the Queen and Prince made tours of inspection during construction, as they frequently did, their visits so excited the workers, the contractor complained, that each visit cost him £20 in lost time.

When it was finally completed in a record-making four months, the building was a tour de force. To construct it had required all the combined efforts and skills of industrial Britain. As much as any of the displays inside it, the Crystal Palace itself reflected British leadership in the industrial age. No other country in Europe could have accomplished it at that time.

Other nations were entering the Industrial Revolution, but none save Britain had yet reached full stride. Belgium boasted a splendid network of railroads, the German mines of the Ruhr were yielding coal, and the textile mills of Alsatian France were hissing steam. But Belgium was a diminutive country, France had little industry outside her textiles, and Germany was a loose confederation of mutually jealous states. On the other side of the Atlantic, the young United States of America had just begun to stir. Only Great Britain combined firm unity of purpose, efficient use of labor and highly developed capital finance. With these assets, England not only was industrially supreme at home, but she controlled one third of the world's international trade as well.

A fortunate combination of circumstances had put England into the lead. She was blessed with great coal fields in Northumberland and Yorkshire, and iron mines in Cumberland; in 1851 two thirds of the world's coal and more than half its iron ore were mined there. This bounty of coal, combined with a shortage of timber that dated back to the 16th Century, had led Englishmen, at the dawn of the 18th Century, to invent a method of smelting iron ore by burning coke instead of charcoal, at once reducing the cost and accelerating the further development of industry.

At about the same time that English inventors began experimenting with coke for charcoal, one of their countrymen, Thomas Savery, devised a steam engine—the single most important invention of the Industrial Revolution, for it powered all kinds of machines before it became the locomotive in the 19th Century. The first steam engine (called at the time a "fire engine") was designed for use in driving pumps to draw subterranean water from the coal mines. It was a cumbersome and stationary affair, but improvements were made little by little, until in 1769 James Watt, a Scottish maker of instruments, patented a workable engine. By 1800 the steam engine derived from Watt's model was used not only in mines, but in forges, textile mills and breweries.

Great Britain was further blessed with access to water, both internally and on her coast; therefore she had inexpensive and easy transportation, a ready means of obtaining what raw materials she did not have and of delivering the products she made to all the ports of the world. Nurtured on the water, England had early developed a large merchant navy, which by 1851 handled nearly 60 per cent of the world's ocean-going tonnage. Where the land was not crossed with waterways, England laid an extensive railroad network.

In addition to the blessings of nature, England had certain social and political advantages over her neighbors on the Continent. Her enterprise had freed itself earlier of the guild system that prevailed in the rest of Europe up to the end of the 18th Century; until that time Europe's guilds had dictated designs and prices, inhibited innovation and restricted the movement of workers. Unlike Continental society, England's was open; younger sons of English peers were not lords, but commoners, who, having to fend for themselves, often went into business, thus revitalizing it with new talent. On the Continent, however, social pressures dissuaded French and German nobles and their sons from entering commerce, and business there remained for generations within the same bourgeois families. This, like the guild restrictions, inhibited innovation.

Politically, England was a unified whole that had no internal boundaries. There were no tolls between its regions, as in France. And England was a nation at peace; for the most part she held aloof from foreign wars and thus did not have to divert her energies into maintaining armies, or deplete her manpower or despoil her land, as did so many nations of the Continent.

Finally, and perhaps most important, England had accumulated great funds of capital, which in turn were the result of a highly developed system of finance that had accompanied industrial growth. Banks had been proliferating since the 17th Century and were proliferating still. They offered credit, much of it on a long-term basis, thus in effect creating capital where there had been none. They had a national network of loans and payments, which enabled one region of the country to draw on another for money. At the head of this sturdy network stood the Bank of England, which reigned over the international world of finance.

The exhibits at the Crystal Palace made British eminence perfectly clear. They occupied the entire west wing of the building and filled half its million square feet of floor space.

First came the raw materials on which English industry was based, and these were displayed in profusion. The principal exhibits in this category were coal and iron ore; a single block of coal weighing 24 tons and standing about the size of a modern limousine loomed proudly outside the Exhibition building. There were also agates from the Isle of Wight, used in mortars and pestles for chemistry and as tools for burnishing. There was gypsum from Somerset, used in plaster of Paris, in tiles and building plaster, as fertilizer and as filler in paper and textiles. There was granite from Argyleshire,

used in buildings and monuments, and lead from Perthshire, used in pipes, cable sheaths and type.

The most popular British exhibits stood in the deafening hall of machinery. Here were the locomotives and marine engines, the hydraulic presses and power looms for which British industry was famous. On display were a 31-ton Great Western locomotive that could travel at the astonishing speed of 60 miles an hour, and an Applegath & Cowper printing press that turned out 5,000 copies of the *Illustrated London News* in an hour. For the proud English visitors, it was an exhilarating experience to stand in the midst of the hissing steam and thundering engines that bespoke their national progress.

As the British were a seafaring people, marine engines held particular appeal for them—the more so because the exhibits made clear the revolution that was taking place in marine history. A model of Watt's 1785 steam engine of one cylinder and 40 horsepower was displayed alongside the latest marine engine of four cylinders and 700 horsepower. As sail was giving way to steam, so iron ships were replacing wooden ones, and these substitutions resulted in a spectacular increase in speed and tonnage. A weekly passenger steamer service between Liverpool and New York had already cut the time of a transatlantic voyage to 12 or 13 days; and though the New England Yankee Clippers, given favorable winds, could still make a faster voyage, they were soon to lose out to the dependable steamships. By the year of the Great Exhibition, Britain had 168,000 of her nearly 4,000,000 tons of overseas shipping under steam. She had no competition.

Just as the British steamship ruled the sea, so the British steam locomotive dominated land transportation, and one giant engine after another indicated as much to visitors at the Crystal Palace. The steam locomotive, like so much else of the

Industrial Revolution, had been of British inspiration. Men had been trying for years to devise a "traveling engine" that could pull a load; then in 1829 George Stephenson, a collier's son who had worked as a cowherd in his boyhood and never learned to read until he was 17 or more, succeeded in doing so. Stephenson's engine would run at only 16 miles an hour, but it nevertheless provided the model for the engine of the future, and it opened an era when a journey from Edinburgh to London, which had taken 44 hours by stage coach, could be accomplished in 12. In 1851, scarcely more than two decades after Stephenson's achievement, over 6,000 miles of track had been laid in Great Britain —more than three times the mileage in France and nearly twice that in the German states.

Britain's triumphs in the production of marine engines and locomotives followed her already established lead in the textile industry. Cotton, wool, silk and linen were still her major exports and the basis of her wealth, and these were lavishly displayed at the Crystal Palace. So were the machines that made them possible. The first was the spinning jenny, invented in the 1760s, when James Hargreaves, a carpenter of Blackburn, devised a contraption by which a spinster could make 30 threads with no more labor than she had previously expended making one. Another one was the water frame, invented by Richard Arkwright, a barber who took a fancy to the process of spinning and sought the aid of a watchmaker to help him rearrange and improve the parts of Hargreaves' jenny so that the threads would emerge uniformly fine or hard. Still another was the power loom, the invention of Edmund Cartwright, an obscure clergyman in Lincoln, who had the idea of applying powered machinery—already in use in mining—to cotton weaving. Finally, there was the mule—probably so called because it was a cross between Hargreaves' jenny and Arkwright's frame—invented by the

only man of the lot who began as a spinner: Samuel Crompton, whose irritation over the defects of the jenny, with its stationary spindles, led him to devote five or six years to devising a movable carriage that remedied the jenny's faults.

The catalog of wonders was seemingly endless. Most of the marvels were British, but there were more than 6,500 exhibitors from other countries of the world, and these showed the varying degrees by which other nations were entering the Industrial Revolution. Only Belgium's rate of development compared with that of Great Britain. Belgium had rich deposits of coal, iron, zinc and marble, and 10 per cent of her small population was engaged in the production of chemicals, iron machinery and linen and woolen textiles. Besides her raw materials, she displayed safety lamps for miners, quantities of lace and books, and some catskins dyed to look like sable. Even in the 19th Century there was a market for inexpensive imitations of the ornaments flaunted by the rich.

France was industrially behind Belgium. Her manufacturing centers were not yet complexes like those in Britain, but rather concentrations of fine craftsmen and skilled artisans. The Industrial Revolution had been slow to develop in France, for a variety of reasons. Internal tolls on rivers and between provinces made the distribution of goods an expensive undertaking for the merchant, and the Napoleonic Wars at the beginning of the century had set the country back by straining her economy and depleting her manpower. Though France won an Exhibition award for a turbine that required about one sixth of the space needed for earlier water wheels and could therefore be used in shallow rivers, or in those where the tides varied, and though she showed some portent of the future with Louis Daguerre's new photography machine, she lagged behind in heavy industry and in the development of railroads.

THE CONSORT'S TICKET *to the Great Exhibition is designated
"Number 1" and bears the signature of Prince Albert, the fair's
chief promoter. Actually, of course, he needed no ticket to enter.*

In products for the luxury market, however, France had no equal, and of these there were extensive displays at the Crystal Palace. From Paris, Sèvres and Limoges came porcelain, crystal, silver and jewels; from Aubusson came carpets; from Lyons, silks and tapestries, and from Grasse, fine perfumes.

Germany, which before the end of the century was to outstrip all other Continental nations in manufacturing, was in 1851 divided into 39 small states that were only beginning to make industrial progress. German exhibits, like those of France, were chiefly the products of craftsmanship—Dresden china and Bavarian porcelain, sculpture and armor. Internal trade, however, unlike that between the provinces of France, was free; a railway system was beginning to fan out from Berlin; and there were at the Crystal Palace exhibits of a few products that were to be of vast importance in the near future: the insulated telegraph wires devised by Werner Siemens, which were soon to make possible the laying of the transatlantic cable, and a six-pounder cast steel cannon made by Alfred Krupp of Essen. The cannon was awarded a gold medal and drew crowds of viewers—but only as a curiosity. In 1851 Europe was still casting cannon in bronze or iron. Krupp could find no buyers for his invention, for few guessed that it was to revolutionize warfare.

Next after Great Britain in variety of entries, but far behind even Germany in industrial production, came the Austrian Empire—a sprawling assortment of states and principalities that spilled across the mountains and plains of the Continent, from the Alps of northern Italy to the forests of Poland. Some of these territories were among the richest lands of Europe in natural resources, but their economy was principally agricultural, their wealth almost entirely in the hands of the aristocracy, and the Industrial Revolution had scarcely penetrated. Austria's most interesting exhibits therefore were the products of painstaking craftsmanship that survived from medieval times, and they were meant for aristocratic consumption—elaborately fashioned glass, porcelain and silver from Bohemia and Vienna.

The Americans, optimistic and boldly self-confident as always, had made the mistake of asking the directors of the Great Exhibition for more space than they could fill. Consequently, their displays were punctuated at regular intervals by pyramids of soap and mounds of dental powder. Most of the American offerings were highly utilitarian, and many were homely—Indian maize, artificial teeth, bankbooks from the City Bank of New York, India rubber products such as lifeboats and pontoons from Goodyear of New Haven. Two of the American exhibits, however, portended the industrial genius of the young country: half a dozen rifles that introduced to England the system of mass production of interchangeable parts, and Cyrus McCormick's reaper, which, though it had been patented as early as 1834, was new enough to most visitors at the Crystal Palace to be one of the sensations of the Great Exhibition. The reaper was soon to revolutionize agricultural methods and production in Europe, as it had already begun to do at home.

From other countries came products showing no signs of the Industrial Revolution: delicate black lace from Barcelona, fine blades from Toledo, polished diamonds from the Netherlands, redolent cigars from Havana, intricate music boxes from Switzerland—and all of them wrought by hand. There were many exotic trifles: pearl oysters from the Indian Ocean, musk from Turkey and rose-water from Tunis.

Whether the exhibits displayed technological advance, exquisite craftsmanship or the riches of the earth, they laid before the eyes of six million visitors a panorama of wealth such as had never been seen before, and that in itself was a triumph of communication and transport. In one place, and for the first time, were gathered together the works of mankind from all over the world.

Nor was it all ostentation. The Great Exhibition stands as a historic national event, and one of far-reaching effects. Some of its profits were put to promoting knowledge of science and its application to industry; some to founding the collection of fine and applied art that was the start of today's Victoria and Albert Museum. Moreover, the fair stimulated international commerce; indeed, it fore-shadowed the era of free trade that was soon to become the dominant principle of 19th Century economic policy.

The majority of those who visited the Crystal Palace, from Queen Victoria to the smallest boy with a shilling ticket, regarded the machine and its products with wondering reverence; they saw machinery as miraculous and hallowed by the light of progress. There were, however, a few dissenters. The art critic John Ruskin carped that the exhibition hall was "neither a palace nor of crystal." Charles Dickens observed, "There's too much. I have only been twice. . . . I have a natural horror of sights, and the fusion of so many sights in one has not decreased it." Others mourned that commerce had become a religion and the factory a shrine. A Swiss writer, Henri Frédéric Amiel, expressed the opinion that an age of mediocrity had begun. "The useful," he wrote, "will take the place of the beautiful, industry of art, political economy of religion, and arithmetic of poetry."

It was beginning to dawn on some men that not all that was wrought by the Industrial Revolution was magic, and thoughtful observers perceived in its triumphs threats to the traditional values of society.

The writer Matthew Arnold, in his poem "Dover Beach," warned:

> . . . the world, which seems
> To lie before us like a land of dreams,
> So various, so beautiful, so new,
> Hath really neither joy, nor love, nor light,
> Nor certitude, nor peace nor help for pain;
> And we are here as on a darkling plain
> Swept with confused alarms of struggle and flight,
> Where ignorant armies clash by night.

In Rome Pope Pius IX saw forbidding threats to religion in the increasing attention to material progress. Dickens in London, Dostoevsky in St. Petersburg, and Hugo, exiled from France, saw threats to human dignity. And Karl Marx was busy in the British Museum gathering material for *Das Kapital*, in which he would urge the overthrow of capitalistic society. Before long the ideas of these men and others like them were to come to the fore in various ways, and with various results.

The darker implications of the age, however, lay hidden from most men in 1851. Most of them shared Prince Albert's optimistic faith in material goods, and in man's ability to triumph with them. Most of them reveled in the Age of Progress as a time of expectation and adventure. For most of them there seemed to be no machine man could not make, no task his machine could not perform, no activity he could not master. The conquest of the material world had begun, and few doubted that it would be to man's unqualified benefit.

THE CRYSTAL PALACE, *with its expanse of glass and airy elegance, was a radical break with the heavy stone architecture of the mid-Victorians.*

THE WONDERFUL FAIR

London's Great Exhibition of 1851 was conceived to show the progress that mankind was making in the 19th Century, and it brought together under one roof displays from all over the world: machinery, manufactured goods, sculpture, raw materials—all the fruits of man's expanding industry and soaring imagination. On opening day, May 1, half a million spectators eagerly thronged Hyde Park to catch a glimpse of what one Victorian poet called "that world's wonder-house —the Crystal Palace."

For months Londoners had talked of little but the fair. They were proud to be hosts to the world's first international exhibition and to subscribe to its noble ideals—peace and industry. They were equally fascinated by the building itself. In less than five months nearly a million square feet of glass had been fitted into a wood and iron forest of 5,000 columns and girders. Glass suddenly seemed the building material of the future; a humorist wrote, "We shall be disappointed if the next generation of London children are not brought up like cucumbers under glass." Word had leaked out about the wonders the structure would house: printing presses and preserved pigs; locomotives and "an alarm bedstead" that could hurl the sleeper out of bed into a cold bath at any given hour; French perfumes; a Prussian stove in the form of a knight dressed in full armor—in short, so the proud Victorians believed, all the highest achievements of human ingenuity.

THE DAZZLING OPENING

Queen Victoria arrived at the Crystal Palace promptly at noon on opening day. She stood on a raised dais under a temporary canopy *(left)* to open the fair and later called it "the happiest, proudest day in my life."

Rarely had Englishmen basked in such splendor. Lush palms fluttered against colorful carpets hung from balconies and water splashed from a 27-foot cut-glass fountain. In the packed galleries women in striped silks and satins and men in formal attire peered through opera glasses at the dignitaries below.

Alarmists had insisted that the fair was sure to attract anarchists who would assassinate Victoria and Albert on the first day and proclaim England a "red" republic. In fact there was no violence at the exhibition during the 23 weeks it was open. The only miscreants were 12 pickpockets and 11 persons caught removing minor exhibits. Among the small number of unclaimed lost articles were three petticoats and two bustles mysteriously left behind.

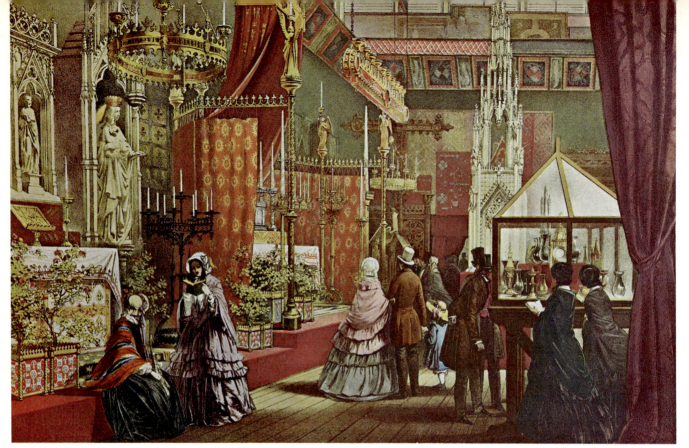

THE OLD WAYS, *represented by a medieval display, are viewed by ladies in full skirts and gentlemen in long coats. Most of the objects shown were church furnishings, such as the tabernacle (left) and the jeweled vessels in the glass case.*

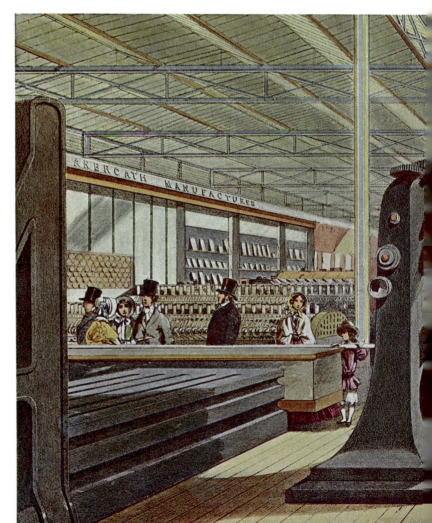

NEW MACHINE TOOLS *invented by Joseph Whitworth dominated the moving-machinery section. In the right foreground is his lathe for shaping railway wheels, an innovation that was important in the expansion of railroads.*

EUROPE'S PAST AND FUTURE

The British exhibits, which filled half of the Crystal Palace, reflected a subtle conflict between the old and the new that was troubling all 19th Century Europe. While people everywhere were looking back sentimentally to the elegance and grace—and above all the simplicity—of other ages, they were also looking eagerly ahead to the exciting new era of the machine. The stronghold of the Old at the fair was the Medieval Court (left), where British factories showed church ornaments, altarpieces and furniture manufactured in the Gothic style. Although many Englishmen professed to admire the ornate grandeur of the room, they soon hurried on to the shiny new hardware and machinery sections.

There factory workers jostled through the biggest crowds at the exhibition to admire the machines and products that they themselves had made —everything from a knife with 80 blades to the newest express locomotive. Steam hammers, hydraulic presses, huge power looms, engines to pull threshers—the whole exploding world of industry and agriculture was on parade. Occasionally history and technology joined to produce such objects of dubious value as an Elizabethan sideboard made of hard rubber. But it was in the gleaming beauty of the machines that the future could be seen.

THE RICHES OF THE EMPIRE

From Britain's colonies came a variety of products that seized the imagination of the English public and inflated its pride. Canada sent furs, a shining new fire engine, bales of hides, casks of maple sugar, Indian regalia, a variety of sleighs and a birch-bark canoe that could hold 20 men. From Australia came hats that convicts had made from the leaves of cabbage trees.

The most fascinating displays were from India *(left)*. Victorian ladies were enraptured with the carved ivory furniture, and the fabulous Koh-i-noor diamond (caged in a burglar-proof contraption of gilded steel and said to be worth more than the rest of the fair's contents put together), and cashmere shawls so delicate that they could be drawn through a wedding ring and so intricately woven that each one took years to make. A fair official explained to the Queen how a British factory in India produced opium—but he failed to mention that the lucrative drug had turned millions of Chinese into addicts.

A CANADIAN CANOE *used by trappers draws the attention of sightseers. In the foreground is a fur-laden sleigh; at the right are barrels of flour.*

AN INDIAN HOWDAH, *complete with a fringed awning, tops an elephant draped with gold and silver trappings. To display these exhibits, fair officials borrowed a stuffed elephant from a museum.*

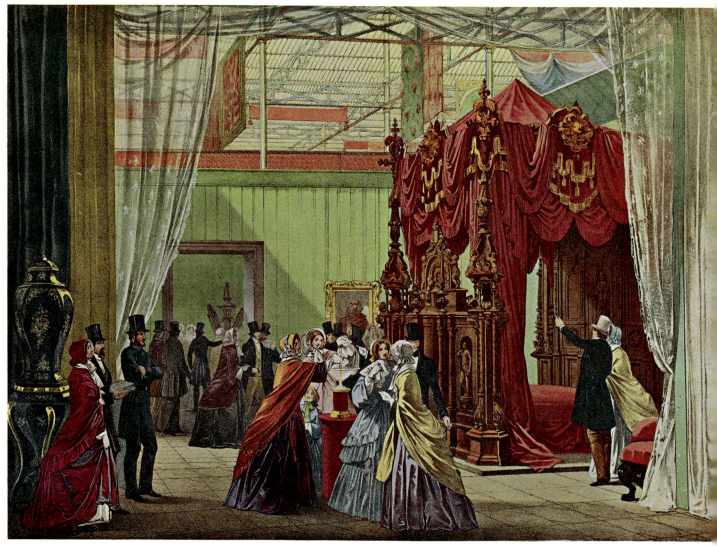

AN AUSTRIAN BED *with an elaborate canopy towers over visitors; in the foreground ladies cluster to dip handkerchiefs in a fountain of eau de cologne.*

THE CONTINENT'S EXQUISITE GOODS

From all over Europe came luxury goods that won the admiration of British consumers and the reluctant envy of local manufacturers. English ladies now realized to their chagrin that the Belgians made more intricate ruffles, the French produced more costly plates and the Austrians designed more imposing beds and bookcases. The superior workmanship of the elegant firearms from Spain and the rich carpets and ornate, gilded bric-a-brac from France was obvious when these handsome articles were compared with their English counterparts.

Fair officials had forbidden exhibitors to sell any goods, but several Continental firms quietly ignored the rules and took orders from visitors—including Queen Victoria, who bought Austrian toys for her children. The huge success of the foreign exhibits jolted English manufacturers out of their smug self-satisfaction, and more than one realized the truth of a line from a contemporary poem by P. J. Bailey that "what England as a nation wants is taste."

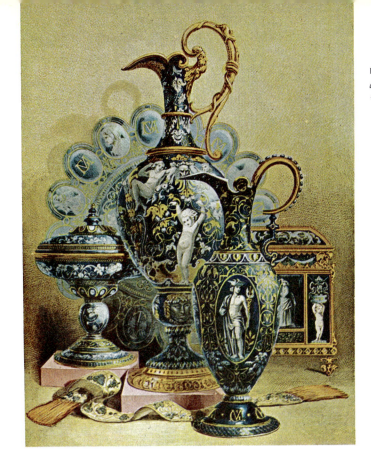

ENAMELED PORCELAIN (left) from the French government factory at Sèvres delighted Victoria with its "taste and execution" that "gave one a wish to buy all one saw." She bought cratefuls.

FRENCH CHINTZ demonstrated what the fair's catalogue praised as "peculiar and graceful indications of artistic feeling." But the practical housewife liked it because it was cheap and tasteful.

SPANISH PISTOLS rivaled the American Colt six-shooter for the attention of gun collectors. Some visitors noted with irony the presence of guns at a fair whose theme was supposed to be "Peace."

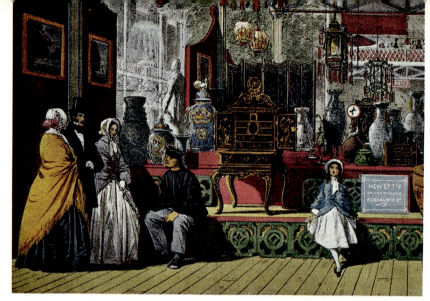

THE CHINESE EXHIBIT *provided carved jade, lacquered cabinets—and a place to sit and rest.*

PRODUCTS OF THE WORLD

For fair officials, some of the most troublesome exhibits were those from distant lands. The Russian display was almost six weeks late; the ships bearing it had to wait for Baltic ports to thaw. The Chinese, who had just lost the first round in their long struggle to stop Britain from sending opium into their country, refused to submit an exhibit, a crisis that led fair officials to borrow Chinese curios from English importers. The Bey of Tunis sent so many odd groups of products that it was impossible to classify some of them, such as the "two scissors used in the red cap manufacture, and some siwak used by Moorish women for whitening their teeth."

One of the American exhibits was a statue called *The Greek Slave (right)* by Hiram Powers. The statue, of a modest, undraped young woman, was particularly pleasing to Victorians because it had a bit of built-in technology: it could be seen from every viewpoint when a crank was turned, making it revolve slowly on its pedestal. It was easily the hit of the show.

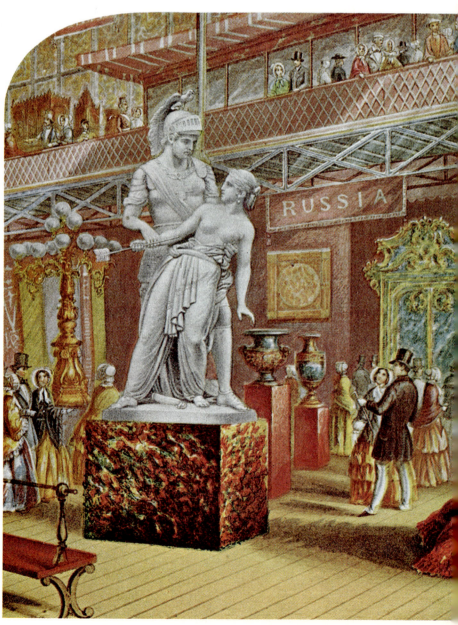

STATUES OF THREE NATIONS *are shown together in this contemporary painting: (left to right) a*

A TUNISIAN BAZAAR, *set up around a nomad's tent draped with lionskins, displayed a variety of outlandish merchandise, all sold (illegally) on the spot.*

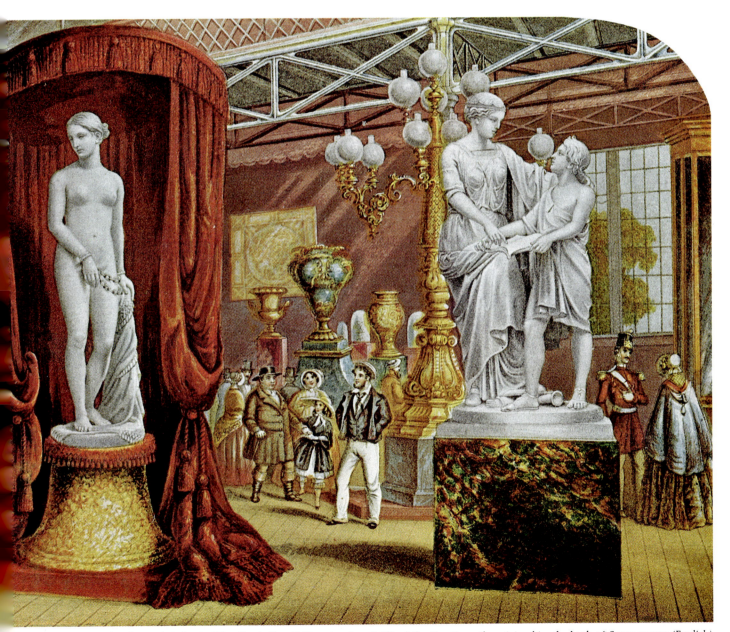

Christian hero and a heathen girl (Italian), "The Greek Slave" (American), and Alfred the Great's mother giving him the book of Saxon poems (English).

2

THE BLESSINGS
OF SCIENCE

Underlying man's unquestioned faith in progress at the time of the Great Exhibition in 1851 was a growing knowledge of science. Science was producing powerful new ideas that were not only yielding additional information about nature, but were actually creating entirely new fields: bacteriology, atomic physics, genetics, psychology, anthropology, sociology. Science had even invented a new method of invention: applied science. All of these developments fostered a faith that through the discovery and application of new scientific knowledge, man could bring the world and himself under his own control and achieve an earthly paradise. For many during the 19th Century, science replaced religion and philosophy as a tower of hope and welfare.

Science had interested thinking men since the days of ancient Greece, but prior to the 19th Century it had been mainly the province of academicians and dilettantes. During the decades following 1850 it became a major intellectual force of Western society.

The Age of Progress was the era when the first comprehensive theories of science were developed and synthesized. It was the era that saw the formation of four of the major concepts on which modern science is based: the idea of evolution; the idea of conservation of energy; the idea of space as a continuum that is pervaded by fields of physical activity (such as electromagnetic fields); and the idea that all action is dependent on the existence of certain basic units—the atom in chemistry, the cell in biology, the quantum in physics and so on. It was the era when science became established in the university curriculum, the era that saw the first associations for the furthering of scientific knowledge, and the era that saw the founding of the elaborate research laboratory, in which the solitary experimentation of the lone individual was superseded by the coordinated work of groups.

The influence of science was not confined to academy and laboratory. Philosophers and politicians, painters and novelists applied its methods and findings to religion, society and art. Writers popularized it in a profusion of books and magazines. Governments teamed with it to provide for public health and sanitation. And finally, inventors made practical application of its theories to push still further the already growing technology.

Today we are accustomed to think of scientific principles as the sine qua non of technology, but this was not always so. Most of the inventions that first tempered man's immediate world were not directly inspired by scientific knowledge; rather, they were the brain-children of ingenious mechanics who employed the empiricism of the factory or the field rather than the theory of the laboratory. They proceeded by trial and error to meet the needs of their tasks or to make them easier, giving little thought to the principles that underlay the results they sought or achieved. Such empiricism had created industry's major source of power, the steam engine, and its offspring, the locomotive and the steamship; it had also created the mechanical reaper and the sewing machine.

Inventors, however, were pragmatists, and so from the outset, although they did not always understand the principles of science, they did employ its methods. They experimented. They weighed and measured, with ever-increasing accuracy. That was how they eventually produced the machine tools that made possible interchangeable parts—and therefore mass production.

It was not until then that the modern phase of the Industrial Revolution could begin. The geometrically shaped parts of a machine—planes, blocks, circles and cylinders—must fit together perfectly if they are to work with speed and accuracy. In the early steam engines of the 18th Century a piston might show a gap of half an inch between itself and its cylinder, and the machine of which it was a part could perform 20 strokes a minute at most and deliver only about five and one half horsepower (about as much as a small outboard motor produces today). By the second half of the 19th Century, length could be measured in the laboratory in millionths of an inch, and enough of this accuracy was car-

ried over to the factory so that the steam engine could soon perform more than 250 strokes a minute and deliver up to 2,400 horsepower.

The development of cheap steel was another high point of the Industrial Revolution; it made possible the building of bridges, railroads, skyscrapers and heavy machines that were beyond the capacity of the more brittle cast iron or flexible wrought iron.

If in the early phase of this period there had been little exchange between the factory and the laboratory, the reason was that scientific knowledge had not caught up with engineering practice. But the rapid expansion, proliferation and refinement of technology toward the middle of the 19th Century was accompanied by advances in the theories of science. A new generation of scientists was coming to maturity, and these men rendered the observations of their predecessors into integrated theoretical structures. They discovered that the methods for probing—and the mathematical formulas for explaining—one branch of science could be applied to another; that physics, chemistry and biology were interrelated disciplines, and that electricity and magnetism were interdependent and unified phenomena.

Meanwhile, scientists and inventors alike began to take notice of the mechanical wonders technology was creating, and to seek to understand and explain how these wonders worked. Both discovered that science and technology were useful to each other. Accurate measurement for precision instruments required a knowledge of mathematics. The empirical inventors began to employ mathematics to estimate the practicability of ideas they sought to incorporate in new machines. Scientists began to examine the steam engine and its transformation of heat into motion, and from these and other studies arrived at the First and Second Laws of Thermodynamics. The first deals with the conservation of energy (it says that energy cannot

A MODEL IN MAUVE, *this turn-of-the-century fashion plate displays a pale purple costume tinted with the first of the new synthetic dyes. The color was so popular that the 1890s were known as "The Mauve Decade."*

by the similarities between the constituents of quinine and a certain derivative of coal tar. In 1849, when thousands were dying of malaria in the British colonies and quinine for treating it was scarce, it occurred to Hofmann that a synthetic drug with the properties of quinine might be derived from coal tar. Hofmann had an assistant, an 18-year-old student, William Henry Perkin, who decided to try to make the synthetic drug that his master had envisioned. He performed one experiment after another and got only a strange black powder.

Others before Perkin had thrown away the results of their failed experiments, but this young man was curious to find out what he had created. He dissolved the powder in alcohol, and the solution turned a brilliant purple. Inspired by the strength of the color, he decided to test it as a dye and dipped some strips of silk in it. When they, too, turned purple, he tested further by laundering the strips in soap and water and hanging them out in the sun, and the color held fast. Perkin had produced the first synthetic dye.

The color purple had been associated with royalty since antiquity, when it could be obtained only from a viscous substance peculiar to the rare shellfish Purpura. In the 19th Century it was extracted from plants, but the process remained difficult and expensive. Realizing the importance of what he had stumbled upon, Perkin left Hofmann and established a factory for the manufacture of his new dye, mauvine; soon he had a thriving industry. The new dye became tremendously popular in the '50s; even penny postage stamps were dyed with it.

Perkin's mauve was only the first of a rainbow of colors that soon flowed from the test tubes of research chemists into the dye vats of the textile industry. Once mauve had been discovered, the newborn industrial chemists turned to making magenta, blue, black and turquoise, substituting manmade chemical compounds for the vegetable and

be created or destroyed; only transformed); the second explains the process by which heat acts as a form of energy.

In chemistry, other scientists discovered that the properties of a chemical compound depend not merely on the atoms it contains (a theory held since the opening of the century, when the English chemist John Dalton established the idea of the atom as the basic unit of chemical combinations), but on the patterns in which these atoms combine themselves in molecules. As they worked out more precise methods for analyzing these patterns and tinkered with them in their laboratories, chemists began to concoct strange new compounds, and to find unexpected uses for them—in other words, they began to make synthetics. Some they arrived at by conscious effort, but some came about by extraordinary accidents.

August von Hofmann, who was director of the Royal College of Chemistry in England, was struck animal substances that had been used for dyeing textiles since the dawn of civilization.

Perkin had launched the synthetic dye industry; the synthetic drug industry was a direct outgrowth. It was found that certain dyes affected some organic substances but not others, and scientists quickly applied this finding to drugs, developing compounds that would strike the agent of a disease while leaving body tissue intact. Other chemists following Perkin's lead devised simpler, cheaper and safer processes for manufacturing old compounds and synthesizing dozens of new ones. They produced explosives for the mining engineer, anesthetics and disinfectants for the physician, even fibers and plastics for textiles and trinkets. In the space of only a few years, celluloid took the place of ivory and bone, rayon the place of silk, and aspirin the place of a host of ancient concoctions.

The influence of laboratory experimentation on the chemical industry was profound, but perhaps even more profound was its influence in the development of electricity. Manifestations of electricity had been known to the Greeks. Their word *elektron* meant "amber"; the Greeks knew that if amber is rubbed it will attract dust and lint by what we now call static electricity.

As late as the 18th Century electricity could be produced only by friction; but electricity thus produced was of no practical value. Then in 1800 an Italian professor of physics at the University of Pavia, Alessandro Volta, made the brilliant deduction that electricity could be continuously generated by chemical action. He piled together in alternating order discs of zinc and silver, sandwiching slices of brine-soaked pasteboard between them, and the two dissimilar metals discharged a stream of electricity. Volta (whose name survives today in the word "volt," a unit of electrical pressure) had fashioned an electric battery. Its current faded and was short-lived, but it was a usable current.

Excitement over the new electricity was widespread, and physicists all over Europe turned to studying it. In 1820 Hans Christian Oersted, a professor in Denmark, discovered while demonstrating the battery that its current would deflect a magnetic needle. In the same year André Ampère, a physicist in France, established the mathematical relationship between electricity and magnetism, and discovered the mathematics of the flow of electrical current. In 1826 Georg Ohm in Germany investigated the resistance of various metals to electricity and discovered Ohm's Law, which describes the relationship between resistance, voltage and current.

Of all the scientists who added to the understanding of electricity, none made a more perceptive and significant contribution than the Scottish physicist James Clerk Maxwell. Maxwell's work was pure theory, not laboratory experimentation, and much of it waited a quarter of a century or more to be proved and put to use, but in 1864 he pulled together the various threads of knowledge pertaining to electricity and magnetism and demonstrated their exact mathematical relationship.

Other scientists had arrived at explanations of physical phenomena by observation; many had constructed apparatus for making things happen, and some, such as Ohm, had used mathematical equations to express their conclusions. But Maxwell took a new stride forward. Applying mathematical equations to what was then known about electric and magnetic fields of action, he postulated the existence of the waves in electromagnetic fields that are now known as radio waves. With further mathematical formulae, Maxwell went on to theorize that these electromagnetic waves travel at the speed of light and thus that radio waves and light are different manifestations of the same phenomenon. His theory was later borne out in the laboratories of other scientists. What Maxwell did was to determine, by use of mathematics, the existence

SCIENTIFIC EXHIBITS *like the one advertised in this poster were held in many European cities and attracted huge crowds. This presentation was on "electro-magnetism as a moving power."*

in nature of physical phenomena unseen and unknown, and to predict how they would behave.

Once the theories governing electricity were understood it became possible to put them to practical use. In 1836 an English chemist, John Frederic Daniell, improved on Volta's battery and succeeded in making one that would produce a strong and steady current. A year later his countrymen Charles Wheatstone and William Cooke used Daniell cells to send messages along a wire, and the telegraph was born. The same year, Wheatstone's invention was duplicated (though on a cruder instrument) by an American painter, Samuel F. B. Morse —who also devised the system of dots and dashes, or long and short vibrations, that for many years served as the basis for the international telegraph code.

Other scientists discovered other uses for electricity. By 1840 battery current was being used to plate base metal with gold, silver and copper. The base metal, when immersed in a solution of salt and the finer metal, would attract and be coated by the finer one. Metalsmiths could now make imitation jewelry, platters, bowls and mugs out of inexpensive metals dressed to look like gold and silver, thus making former luxury items available, like the erstwhile royal purple, to the poor as well as the rich.

The chemicals needed to produce electricity by batteries were expensive for industrial use, however, and if further applications were to be made of electricity a cheaper source had to be found. Many scientists, speculating on Oersted's discovery that electricity creates magnetism, wondered if the inverse were true: could magnetism produce electricity? The answer, which led to a major advance in the technology of electricity, came in the early 1830s with almost simultaneous discoveries on opposite sides of the Atlantic—by Joseph Henry, an American professor of physics, and Michael

33

Faraday, an English research physicist. They found that a wire passed through the field of force of a magnet did indeed produce an electric current.

The steam engine was promptly put to work spinning a magnet around a coil of wire—or spinning the coil around the magnet—and the dynamo was born. The first commercially successful dynamo was developed by Zénobe Théophile Gramme, a Belgian. Only a decade after the discovery of the principle, dynamos began to find a score of special uses. In the 1840s they replaced batteries to power electroplating works. Soon afterward they were used to generate light; in 1858 British navigational authorities installed on the southern coast of England a dynamo-powered arc light whose intense beams helped guide mariners through the Strait of Dover. In 1865 an American metallurgist, James Balleny Elkington, developed a process for refining copper—a product much needed by all branches of the burgeoning electrical industry, because it is one of the best conductors of electricity —by still another application of electricity itself. He immersed plates of unrefined copper together with wires of pure copper in a solution of copper sulfate, sent a current of electricity through the system, and succeeded in drawing pure copper off the unrefined plates and depositing it on the wires. The principle behind the process was basically similar to that governing electroplating.

Electricity in 1865 was still, however, limited to special purposes, partly because it had to be generated on the spot where it was used. The credit for converting it into an everyday commodity in everyman's life belongs in great measure to the American inventor Thomas Alva Edison.

Edison is often portrayed as an inspired but unschooled tinkerer who drew on native canniness rather than scientific knowledge—an impression reinforced by his own boast, "I can hire mathematicians but mathematicians can't hire me!" Hire

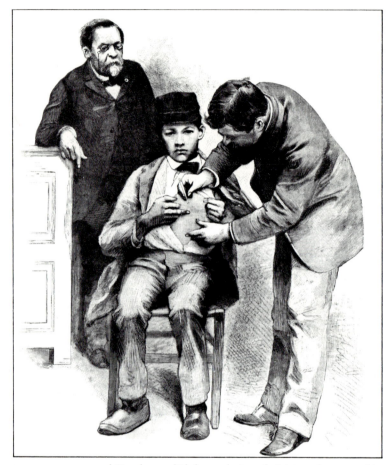

VACCINATING A BOY *bitten by a rabid dog, a doctor administers the new rabies serum developed by Louis Pasteur, while Pasteur himself watches the process anxiously. Rabies had almost always been fatal before the serum, but the boy survived.*

mathematicians he did, however, and because he supplemented his own wits with extensive reading and with the resources of men trained in the rigorous theories and mathematical methods of electrophysics, no one illustrates better than Edison the 19th Century merging of theoretical science with technological ingenuity.

The system that he devised consisted of a new kind of generator that produced a steady current at half the fuel cost of any previous generator, an incandescent lamp to draw on the current produced by this dynamo, and even an electric meter for billing the customers of the system. In the space of 10 years Edison lights were illuminating public buildings, factories and homes in industrial cities around the world.

When the dynamo was first conceived, its inventors used the source of mechanical energy nearest at hand, the steam engine. This was still a relatively crude device, capable through its reciprocating motion of driving a generator wheel at only about 250 revolutions per minute. Soon, however, engineers began feeding the steam jet directly into a turbine; this far more efficient system (still widely used) delivered up to 18,000 rpm.

At about the same time, electrical experimenters turned to another source of energy that had been known to antiquity but largely eclipsed since the coming of steam: water power. Water power had been grinding grain for centuries. But the ancient water wheel had to be located next to a stream; it was not portable like the steam engine, which could operate on any site to which coal could be brought. By hitching a water wheel to a dynamo, however, and thus converting its energy into electricity, water power could in effect by transported over hundreds of miles. In the first decade of the 19th Century, Niagara Falls was a virgin cataract in a northern wilderness; by the last decade of the century it was pouring 15,000 kilowatts of electricity into United States and Canadian power lines. Across the Atlantic, melting Alpine snows were driving dynamos to create industrial energy in coal-poor Switzerland and northern Italy.

The implications of Edison's work are as significant in the social sphere as they are in science and technology. His laboratory in Menlo Park, New Jersey, was the first full-time industrial research organization in the United States, and one of the pioneers in the world. It was a human institution of a new and revolutionary aspect. Prior to the late 19th Century the organized bodies of men that ran the affairs of the world—parliaments, bureaucracies, churches, guilds and corporations—had been for the most part concerned with preserving the status quo, or at least with maintaining stability. Change, when it occurred, was generally the result of special circumstances or haphazard work by individuals. In industrial research laboratories like Edison's, society had for the first time made room for an institution whose permanent, full-time occupation was innovation. Technological, and ultimately social, change, with all they imply in opportunity and peril, excitement and anxiety, began to be built into society with the founding of the research laboratory.

This new institution prospered all over the industrial world, most conspicuously in Germany. The German dyemakers, for example, established research laboratories, sometimes with the assistance of the government, which also fostered university training that would in turn feed the industry new talent. The German electrical industry did the same. The result was that by the end of the 19th Century German technology had outstripped that of England, the mother of the Industrial Revolution and still its leader at the time of the Great Exhibition; of the Nobel prizes awarded in physics and chemistry up to the time of the Hitler era (a time of darkness for both Germany and German

SPOOFING EVOLUTION, *a cartoon of the day portrays a simi-an Charles Darwin explaining his controversial theory of evolution to an ape with the help of a mirror. The work appeared in the "London Sketch Book" in May 1874, captioned by two suitable quotations from the plays of Shakespeare: "This is the ape of form" and "Four or five descents since."*

science), about one third of them went to Germans.

Not all the advances of science had to do with commerce and industry; one of the most striking had to do with human survival itself.

The 19th Century was the first to develop a broad program of public health. Epidemic disease had plagued mankind since the first cities were built. Men had met it with prayer, potions, and sometimes—perhaps prompted by superstition, perhaps by a dim notion that objects once in contact with disease might be infectious—by quarantine or by "fumigations," the burning of objects with which the diseased had been associated. By the middle of the 19th Century, the incidence of typhoid and cholera had increased beyond anything known to historical annals; they had, in fact, increased in proportion with the growth of industrial cities; where the poor, crowded into filthy tenements, with meager supplies of water, and most of that polluted, lived like pigs and died like flies. But the Age of Progress was animated by a conviction that rationally applied human effort could transform the world; and humanitarian men addressed themselves to public health.

The "sanitarians," as they were called, began before mid-century to investigate the living conditions of the workers in England. Their best-known figure was Edwin Chadwick, a lawyer, who in 1842 published a report called "Sanitary Conditions of the Laboring Population of Great Britain." His report laid the foundation for a sanitary movement that began in England and then spread across Europe. In 1848 a cholera epidemic in London gave grim evidence that disease was no respecter of income or class, and the sanitarians began to achieve results: that year the British government created a General Board of Health. London, and then other cities—Paris, Berlin, Munich, New York—began to acquire pure water supplies and public sewage systems.

The sanitarians had begun as empirically as their counterparts in the world of technology. They knew little of the nature of disease; they simply ascribed it vaguely to "filth" and set about wiping it out as best they knew how. With the work of the chemist Louis Pasteur in France and the physician Robert Koch in Germany, however, sanitation acquired a scientific rationale and a specific target: not filth in general, but microbes. Microbes (from the Greek *mikros*, "small," plus *bios*, "life") are living organisms too small for the unaided eye to see. They make many positive contributions to human enjoyment; they are responsible for aging cheese, tanning leather and leavening dough. But they are also the cause of infectious disease. Once Pasteur had established this and grasped an understanding of the nature of the organisms, he and his successors were able to develop vaccines to guard against the deleterious microbes. The subsequent adoption of immunization, together with provisions for cleaner living and working conditions for the poor, almost wiped out epidemic disease in the industrialized world in the space of only a few decades. In 1871 the death rate from typhoid was 332 per million; by 1911 it was down to 35 per million.

Oddly enough—considering 19th Century preoccupation with material goods—the most profound impact of science on thought came from research that had no technological manifestations whatever: Charles Darwin's theory of evolution, which declared that existing species of plants and animals had evolved into their present forms by a process of natural selection over millions of years. Its corollary implication, that man was not the result of special Creation, but had evolved with, and out of, the animal kingdom, rocked the entire 19th Century world.

Revolutionary as the idea was in its challenge to traditional tenets, developmental change, or evolution, had in fact been accepted in some fields of science and in history. Naturalists of the 18th Century had compiled comparisons of the vertebrate animals and placed them in the same category as man. Geologists understood the significance of unfamiliar fossils found deep in the earth, and had deduced that the antiquity of the earth did not tally with a literal reading of the *Book of Genesis*. The 18th Century chemist Antoine Lavoisier, in recognizing that oxygen is consumed by the body, had implied that life was a process of chemical change. Historians were coming to think of civilization as an unfolding process and to preach that the only way to understand one era was to understand its predecessors. Moreover, the prevailing idea of "progress" itself was an evolutionary concept.

But Darwin performed the monumental task of crystallizing the nebulous ideas in which evolution was implicit and, articulating them in biological terms, upset man's concept of himself and his place in the universe. He ranks with Galileo, Newton and Einstein as an instigator of a major revolution in the history of scientific thought.

Darwin's thesis made four main points: that organisms do not reproduce identical replicas of their kind, but rather produce variations, many of which are hereditary; that nature allows the survival of only those organisms that can adapt to their environment; that all organisms therefore undergo a struggle for existence; and that out of this process comes the survival of the fittest. Many of Darwin's details have since been modified or expanded, but his theory remains at the heart of all modern thinking on the subject.

When the theory was published in 1859, in Darwin's book *The Origin of Species*, it sparked a furious debate. Theologians and churchmen were alarmed, even outraged, by the thesis; they viewed it as a degradation of man and feared a loss of morality as well as religious belief. Darwin himself (and his supporters, who included many of the great scientific minds of the day) regarded the process as a noble one. The elaborately constructed forms of nature, he wrote, "have all been produced by laws acting all around us. . . . Thus . . . the most exalted object which we are capable of conceiving, namely the production of higher animals, directly follows. There is grandeur in this view of life."

Whether praised or scorned, evolution caught men's minds at once. No other scientific idea of the 19th Century had such far-reaching effects on contemporary thought. It was easily understood because it was not highly abstract or mathematical, and it was sensational because it challenged traditional doctrine.

Moreover, it reached all quarters of society. The great scientist Thomas Henry Huxley counted among his most important achievements his lectures on science to laborers in London—and the theory of evolution was included in these. Public education was teaching even the lower classes to read, and literature of every kind was available to all in the new public libraries—another 19th Century innovation.

Theories less comprehensible than Darwin's, but equally portentous for Western civilization, were meanwhile taking shape. By the turn of the century, some of the basic postulates of early 19th Century science were coming under revision. In 1897 Sir Joseph Thomson, a physicist at Cambridge, discovered the first atomic particle, the electron; he thus initiated the new experimental science of atomic physics. Atoms themselves, which James Clerk Maxwell had viewed purely as models "unbroken and unworn . . . perfect in number and measure and weight" since the beginning of the universe, were found to be unstable. Moreover, the breakdown of atoms appeared to be creating the uncreatable. The radioactive decay of uranium was "producing" energy—a phenomenon that the First Law of Thermodynamics held to be impossible.

Even more startling, to the few who could understand them in 1905, were the theories of a young clerk in the Swiss patent office, Albert Einstein, whose equations indicated that the most fundamental aspects of the physical world—time and space—could not be measured in absolute terms.

The impact of these discoveries lay a quarter of a century in the future, but all the foundations of 20th Century science and technology were laid in the Age of Progress. A crude motion picture camera had been invented by 1888. In 1886 an official of the American census bureau, Herman Hollerith, wishing to speed up his job of tabulation, invented a method of punching information onto cards and collating the data with an electric machine. What he had was a crude computer. Radio communication was achieved in 1896, when Guglielmo Mar-

coni, a young electrical engineer from Bologna, devised a machine that would generate the radio waves whose existence Maxwell had foreseen.

Even information had become a mass-produced commodity before the close of the century. The rotary printing press, photoengraving, the linotype and cheap paper from wood pulp turned out volumes of books and magazines, penny cyclopedias and newspapers with circulations reaching into the hundreds of thousands. The newspapers were further aided by the telegraph and oceanic cable, which carried ideas around the globe with the speed of light. The prospectus for *The Daily Express* of London when it began publication in 1900 boasted that "No event can occur in the most remote corner of the earth without *The Daily Express* being placed in immediate possession of its fullest details."

Nor was there any part of the world that did not feel the impact of Europe. Industrial production was drawing raw materials from faraway China and India, Malaysia and Japan. It was spewing out mountains of goods that could not be expended at home and therefore required markets abroad, which in turn gave impetus to conquest. The steamship could deliver European troops, as well as goods, to any land that caught a European nation's fancy. Before the Age of Progress faded, there was hardly a corner of Asia or Africa where European thought and technology had not traveled.

Science and technology not only greatly enlarged man's understanding of the natural world, as midcentury optimists expected they would; together with the resulting confrontation of different classes and different peoples, they were to reshape man's view of society and his economics, temper his religion and his philosophy, and influence his art and his literature as well. In so doing they were to pose in a hundred ways a knotty question: Where was progress taking Western man?

THE EIFFEL TOWER, *shown here half built, rises above other construction at the 1889 Paris Universal Exposition.*

TECHNOLOGY'S MASTER STROKES

In pursuit of visions as bizarre as Gustave Eiffel's spidery, 984-foot iron tower (*above*), inventors and engineers created a new world. It was a technologist's utopia, assembled out of steel girders, illuminated with electric lights, powered with dynamos and gasoline engines, linked by copper wires. Its heroes were men who quickened technology's stream with astounding creations: Edison's incandescent bulb, Bessemer's converter, Roentgen's X-ray tubes, Eastman's camera, Baekeland's plastics, Bell's telephone, Marconi's wireless, the Wrights' airplane. These wonders were every bit as novel as the fanciful mechanical prophecies sketched centuries before by Leonardo da Vinci—only this time they worked.

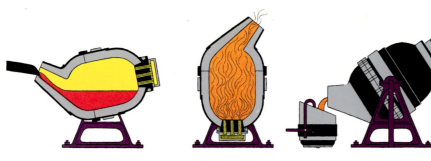

BESSEMER'S PROCESS *converted molten pig iron into steel inside a gourd-shaped furnace, or "converter." The furnace was first charged with the iron (left), then tilted upright while a stream of air was blasted through holes in its base, burning off carbon and impurities in a shower of sparks (center). In about 20 minutes the iron had been transformed into steel and was ready to pour into a ladle (right) to be cast as ingots.*

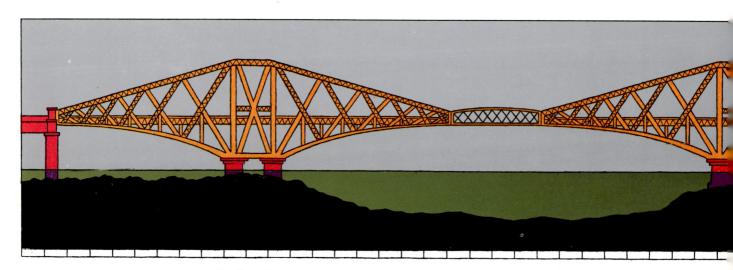

THE OPEN-HEARTH FURNACE *(below), developed by English and French inventors soon after Bessemer's process, produced steel of higher tensile strength that made possible structures like the Firth of Forth Bridge (above). In the furnace, the carbon and impurities in a central pool of molten pig iron were gradually burned off in a mixture of air and flaming, preheated gases that circulated between brick heating chambers.*

NEW METAL: STEEL

In 1856 the English inventor Henry Bessemer developed a revolutionary process of producing cheap steel. Suddenly the age had a metal tougher, lighter and more flexible than iron to put to its most ambitious tasks. Trains now ran faster over durable steel rails, lightweight steel plating gave ships more buoyancy and cargo space, buildings rose higher on skeletal steel frames, and warfare acquired an arsenal of steel artillery and armor. As steel production soared from some 500,000 tons in 1870 to 28 million in 1900, factory cities sprang up at Pittsburgh, Sheffield, Essen and Lille, creating multimillionaires like Andrew Carnegie and the Krupps.

In 1890 engineers convincingly showed off the new metal's strength by erecting a gigantic, all-steel railroad bridge over Scotland's Firth of Forth (below), its 1,710-foot spans for years the longest ever dared by man. Though the artist William Morris called the bridge the "supremest specimen of all ugliness," steel's supple strength had ushered in a new and more functional architecture.

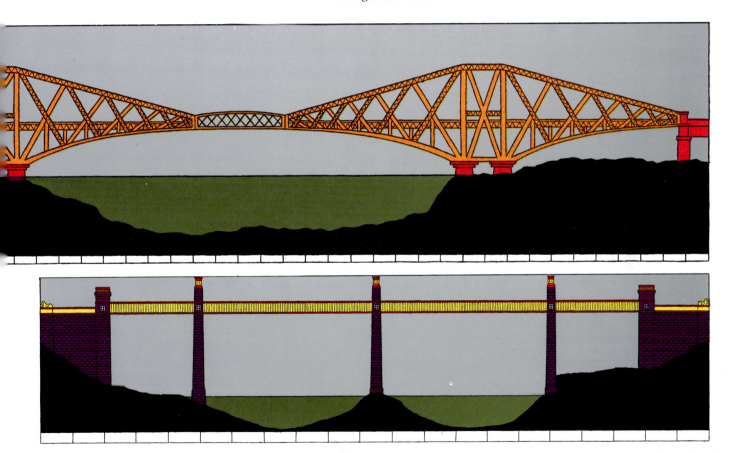

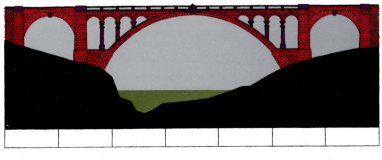

STEEL'S SUPERIORITY over masonry and wrought iron is illustrated in these scaled drawings of bridges. Using high-tensile structural steel, engineers were able to stretch each of the cantilevered spans of the Firth of Forth Bridge (top) across 1,710 feet of open water. Wrought iron, a common bridge-building metal in the late 19th Century, permitted strong spans of only 500-600 feet between piers, as in the Britannia Bridge in Wales (center). Masonry was even more limited: Luxembourg's 1903 Pont Adolphe (right) is the world's second-longest stone bridge, but its main arch spans only about 300 feet.

THE ATLANTIC CABLE *transmitted messages tapped out on conventional telegraph keys (left). Instead of receiving messages as dots* *and dashes, a special receiver (right) translated the faint electrical impulses into light flashes that could be read off a coded scale.*

OCEAN CABLES

In 1866 the steamship *Great Eastern* reached Newfoundland, completing the historic work of laying a telegraph cable that stretched unbroken from Ireland, under 2,000 miles of the Atlantic. On both sides of the ocean it was hailed as one of the century's epic achievements. The cable linked the telegraph systems of the Old World and the New, almost instantaneously creating a revolution in the transmission of news and business information.

Merchants awoke to a world market; with a flicker of a telegraph key they were able to dispatch tramp steamers from distant ports to deliver cargoes bought and sold by wire. News agencies sprang up, transmitting word of human events minutes after they occurred. New undersea cables brought the world network to Australia in 1871 and South America in 1874; in 1902 a British cable spanned the Pacific.

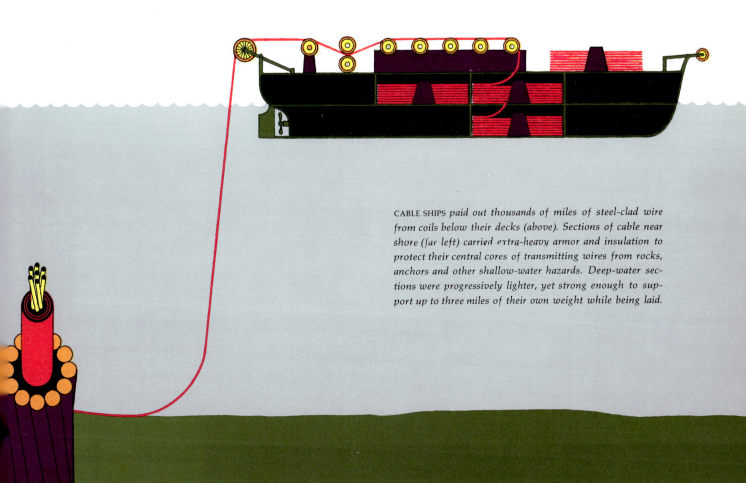

CABLE SHIPS *paid out thousands of miles of steel-clad wire from coils below their decks (above). Sections of cable near shore (far left) carried extra-heavy armor and insulation to protect their central cores of transmitting wires from rocks, anchors and other shallow-water hazards. Deep-water sections were progressively lighter, yet strong enough to support up to three miles of their own weight while being laid.*

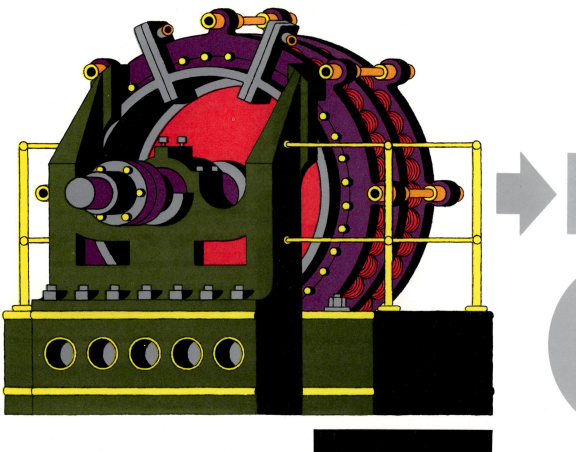

POWER BY WIRE

The electrical era came of age in a blaze of light after the "Wizard" Thomas Edison perfected the incandescent lamp in 1879. Edison demonstrated the new invention at his Menlo Park workshop before 3,000 spectators, and the New York *Herald* splashed the news over its front page: "It makes a Light, Without Gas or Flame, Cheaper Than Oil."

Within a year, Edison was at work designing New York City's first central power station. Built on a slum street where land was cheap, its steam-driven dynamos at times shook the earth and scattered sparks, and its meters overcharged customers. But the system worked, and was widely copied, making current available for a host of new devices.

THE DYNAMO, *diagramed above, created electricity from a coil rotating inside a magnetic field. By the 1890s huge, steam-driven dynamos (top) were feeding current to ever-expanding utility networks.*

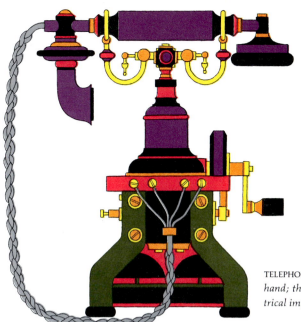

POWER PLANTS *generated electric-ity by using steam or water power in turbines to turn their dynamos.*

STEP-UP TRANSFORMERS *raised the voltage from banks of dynamos for transmission over long distances.*

HIGH-TENSION WIRES *carried cur-rent generated by distant suppliers for areas lacking power sources.*

STEP-DOWN TRANSFORMERS *cut high voltages to levels safe for homes and industrial machinery.*

HOME OUTLETS *tapped the ready supply of energy to power lights, sewing machines, heaters and fans.*

TELEPHONES *had bells cranked by hand; they converted sound to elec-trical impulses, then back to sound.*

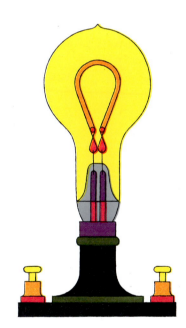

LIGHT BULBS *marketed in 1881 used bamboo filaments held by platinum clips, burned for about 1,400 hours.*

ELECTRIC LOCOMOTIVES *such as this double-ender drew their power from high-voltage lines strung overhead.*

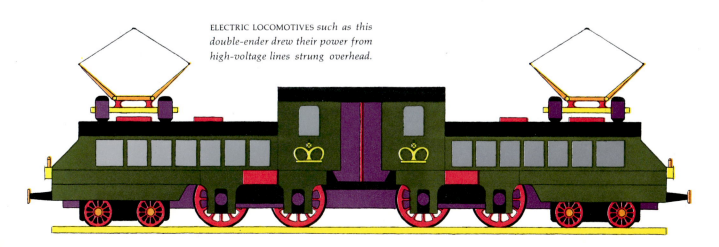

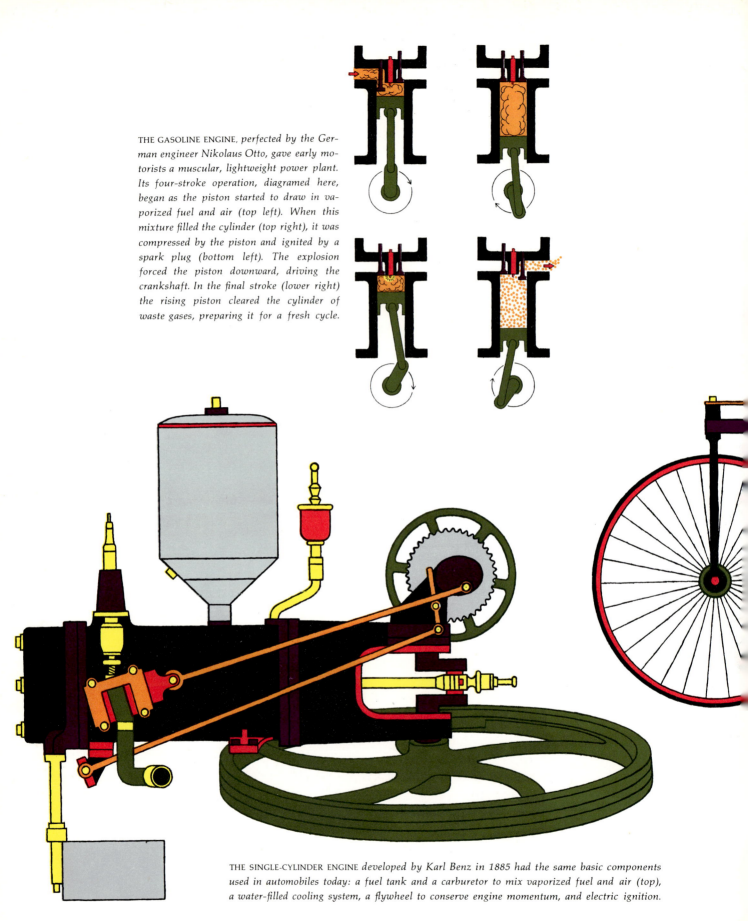

THE GASOLINE ENGINE, *perfected by the German engineer Nikolaus Otto, gave early motorists a muscular, lightweight power plant. Its four-stroke operation, diagramed here, began as the piston started to draw in vaporized fuel and air (top left). When this mixture filled the cylinder (top right), it was compressed by the piston and ignited by a spark plug (bottom left). The explosion forced the piston downward, driving the crankshaft. In the final stroke (lower right) the rising piston cleared the cylinder of waste gases, preparing it for a fresh cycle.*

THE SINGLE-CYLINDER ENGINE *developed by Karl Benz in 1885 had the same basic components used in automobiles today: a fuel tank and a carburetor to mix vaporized fuel and air (top), a water-filled cooling system, a flywheel to conserve engine momentum, and electric ignition.*

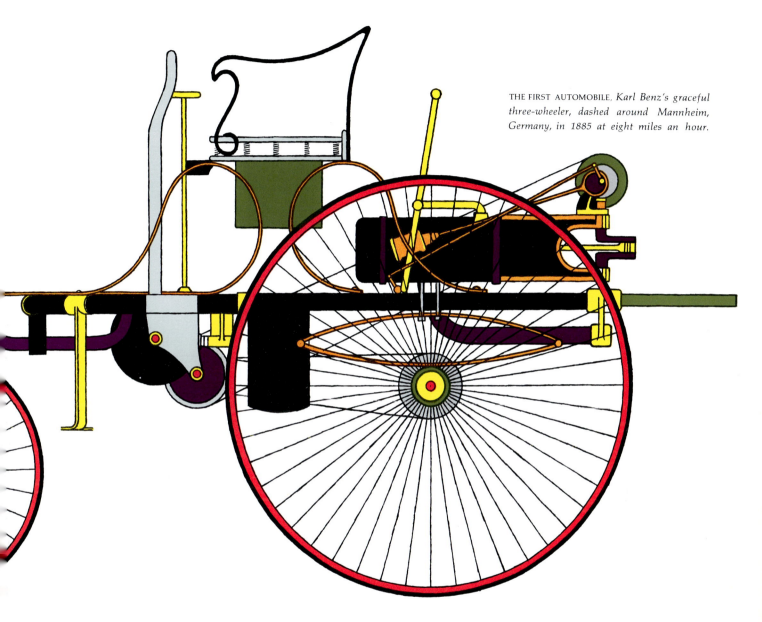

THE FIRST AUTOMOBILE. *Karl Benz's graceful three-wheeler, dashed around Mannheim, Germany, in 1885 at eight miles an hour.*

THE AMAZING MOTOR CAR

The automobile, an ingenious German invention that was to provide mobility for the masses in the 20th Century, began its 19th Century career as a plaything of the rich. Wealthy enthusiasts raced the flimsy contraptions throughout France (which had the best roads) and up Alpine hills—sometimes spending half their time patching Dunlop's new pneumatic tires. For ladies who dared the dusty roads, Paris couturiers advertised the latest in goggles and long coats. In England the Red Flag Act, inspired by stagecoach lines, temporarily held back motor cars by requiring a man to walk ahead of them to warn people with his flag. But in 1896 the Act was repealed, and England's gentlefolk took to the roads. Magazines like *The Automotor* and *Horseless Vehicle Journal* initiated them into the mysteries of shifting gears and filling petrol tanks. The age of the automobile had finally begun.

3

THE WAY
TO WEALTH

Man has lived by trading his possessions for someone else's since the rise of civilization. But the mechanization of European industry in the latter half of the 19th Century, with its unprecedented proliferation of manufacture, brought about an expansion of trade that was more extensive, more complex and more lucrative than anything previously known. It sent men to the farthest shores of the world in search of raw materials and new markets. The manufacture and exchange of goods led to the creation and exchange of capital, which in turn resulted in a vast and interdependent economic network. Finally, the growth of trade created a gushing fountain of wealth that looked to many as though it would never run dry.

The increasing complexity of trade had effects on geography, on political power and on social structure. It resulted in the forging of major national powers, sometimes sharing common interests, sometimes at odds with one another, but among them capable of determining the destiny of the rest of the world. It spawned gargantuan and immensely influential political and social institutions that were to bring about the most rapid and widespread transformation of mankind ever seen.

The causes behind the rise of European commercial supremacy were as diverse and complex as its effects. But a major clue to understanding this phenomenon lies in the remarkable growth of population that accompanied the Industrial Revolution. In 1815, a generation after the initial phase of industrialization had begun, the population of the entire European Continent stood at about 200 million, scarcely more than the population of the United States in the mid-1960s. During the rest of the 19th Century, and into the first decade of the 20th, population increased at the rate of .75 per cent per annum—a rate of growth never before generated by so large a group of people for so long a period. By 1914 the Continent's population had more than doubled (Europe's 460 million now constituted a quarter of the world's inhabitants), and emigrants from Europe to other parts of the world, plus their descendants, numbered an additional 200 million. With its influence thus dispersed, European civilization could scarcely fail to imprint itself upon the world.

THE WEALTH OF EUROPE, *reflected here by the paper currencies of many nations, increased enormously with the mechanization of industry; eventually it created a network of capital that dominated the world's economy.*

Europe's large population both drew on and nourished the Industrial Revolution. With more mouths to feed, backs to clothe, heads to shelter and tastes to satisfy, industry thrived on widened markets. At the same time, the growing population provided the labor force needed to man the factories and to enable industry to grow ever faster.

At first the focus of the Industrial Revolution rested almost entirely on England, but within two decades of the Great Exhibition of 1851 Continental nations had begun to take new strides in manufacturing. The result was a mounting challenge to Britain's lead in industrialization. At first the Continental nations learned from Britain, drawing on its skills and its capital, its workmen and its engineers. But soon these countries were proceeding on their own initiative to build machinery, to manufacture goods and to construct railroads, steamships, bridges and canals to transport and sell the new manufactures abroad.

The advance of industrialization did not, of course, proceed uniformly; it moved at different rates, depending on the resources and on the social, political and economic structures of different nations. The result was the emergence by 1870 of a Europe divided into an inner zone and an outer zone. The inner zone comprised the industrialized nations or regions: Great Britain, Belgium, Germany, France, northern Italy and the western areas of the Austrian Empire. In addition, in every sense but the geographical, the northeastern United States could be considered a part of the same group. Not only did these regions possess virtually all the wealth of Europe in the form of heavy industry and extensive railway mileage, but within their borders were to be found most of the scientific laboratories whose findings contributed to both the rapid advances of industry and the improvement in health conditions essential to a strong working force. Moreover, the nations of the inner zone had parliamentary and constitutional governments, they were the sites of most of the humanitarian and reform movements of the day, and within them literacy was becoming universal.

The outer zone was agricultural rather than industrial; it included most of Spain and southern Italy, most of Ireland and all of Europe east of Germany and Austria. This zone was dominated by wealthy landowners, who were served by a poor and illiterate peasantry whose members had a short life expectancy and little hope of self-improvement. After 1870 the nations of the outer zone came to live by selling the products of their soil to the industrial nations, but they never made enough thereby to purchase any great quantities of manufactured goods in exchange.

VISION OF A TRAIN, *this sketch by Friedrich List, a German econ-
omist, was made before there were railroads in Germany. The
flatcars bearing carriages anticipated modern "piggybacking."*

Beyond the two zones of Europe lay a third zone of the world: the great tracts of Asia and Africa, most of them undeveloped and all of them fated to come under the sway of Europe after 1870. They offered abundant supplies of metals for coinage, jute for making the bags in which manufactures were packaged, and palm and coconut oil for soap, paint and lubricants. And they yielded rubber, which was used for tires to shoe the bicycle—a 19th Century invention that took Europe and the United States by storm; it became at once a fad, a sport, an important means of transportation and a major industry. In addition to raw materials for manufacturing, the European nations bought much of their food abroad: mutton from Australia, beef from South America, sugar from Cuba, spices from the East Indies and tea from China. In the last 30 years of the 19th Century, there was hardly a port in the world where European merchant vessels could not be found, either picking up raw materials to take home or unloading finished manufactures. The volume of export-import trade reached billions of dollars.

As they acquired surplus wealth through industry and trade, European nations added another important commodity—money—to their list of exports. As capital investment, European money built railroads and bridges, founded companies to do more trade—and paved the way for later colonial imperialism. Britain remained to the end of the era the greatest exporter of capital; between 1880 and 1914 its investments abroad nearly tripled, until they reached some $20 billion and constituted one quarter of the nation's total wealth. But in the same period French investments overseas more than doubled, reaching almost nine billion dollars, and German investments more than quadrupled, reaching six billion dollars.

None of this expansion of trade and capital would have been possible without the growth of a network of communication and transport. By 1871 the tele-graph made it possible for an entrepreneur in London or Paris to give orders as far away as Calcutta and Sydney; humming wires rapidly brought him the latest prices of wheat in Minneapolis, Danzig or Buenos Aires, and he could speedily execute his orders to buy or sell accordingly. In 1874, 22 nations agreed by treaty to establish the Universal Postal Union, a new institution that ended a confusing array of rates and weight restrictions, and that provided the first guarantee of uniform mail service throughout the industrial world.

The expansion of rail service had an even more telling effect on trade. In 1850 world railway mileage stood at 23,200, about 25 per cent of which was in Britain. By 1900 the figure reached 491,700 miles; rails extended as far as Constantinople and Vladivostok and penetrated parts of Australia, Latin America and India. By 1914 freight trains were rumbling to every important trading area in the world.

Governments sometimes took a hand in building railroads, but for the most part the new lines were the work of enterprising private businessmen. One British contractor, Thomas Brassey, in the space of a few decades laid at least 8,000 miles of track in Britain, France, Holland, Prussia, Spain, Italy, Canada, India, Australia and South America. At one time he simultaneously employed some 75,000 men on five continents, and on one of his projects the men spoke 11 languages. Brassey was notable not only for the vastness of his operations, but also for his concern for his men's welfare; realizing that they could not work on empty stomachs, he saw to it that they were well fed.

With the same vigor with which Brassey laid rails, an imaginative French engineer, Alexandre Gustave Eiffel, built bridges all over Europe. Eiffel is generally regarded as the father of iron and steel construction in France; much of his work is still in use, giving testimony to his skill and versatility. Among his most famous bridges are the 525-foot span over the Douro River at Oporto, Portugal, and the Garabit railroad viaduct over the Truyère River in south-central France. This bridge, a great arch that crosses a 406-foot-deep ravine in a single span of 540 feet, was an original and daring construction and remained for many years the highest bridge in the world.

In recognition of Eiffel's many accomplishments, the directors of the Paris Exposition of 1889 commissioned him to build a structure that would symbolize the fair for all visitors—a practice imitated by fairs ever since. Fairs and their projects have a way of causing controversy, and Parisians, complaining that Eiffel's soaring 984-foot tower marred the beauty of the city, derisively dubbed it the "junkman's Notre Dame." Distinguished Parisians, including the composer Charles Gounod and the novelist Alexandre Dumas the younger, signed a petition to protest the Eiffel Tower; Guy de Maupassant, the writer, left the city for a while to express his disgust. Nevertheless, two million fairgoers ascended its girders in its first year, and many more millions have been doing so ever since.

What railroads and bridges did to extend land transport during the Age of Progress, steamships and canals did for water-borne transport. One of the heroes of shipbuilding was Isambard Kingdom Brunel, an English civil engineer who at 27 was already successful at building railroads before he took an interest in steam navigation. He subsequently constructed three pioneering steam vessels: the *Great Western*, the first steamship designed specifically for transatlantic voyages; the *Great Britain*, the first ship made of iron and the first ocean vessel to use the screw propeller; and the celebrated *Great Eastern*, a mammoth but graceful ship nearly double the length of most ships of her time and able to carry 4,000 passengers—nearly twice the number carried by the *Queen Elizabeth*, the largest passenger ship in service today.

The *Great Eastern's* size represented a practical effort by a brilliant engineer to provide the space needed for profitable cargoes and the fuel supply needed to voyage around the Cape of Good Hope to India and Australia. But before the *Great Eastern* was put afloat, other plans were afoot that were to make her daring size a handicap and consign her to the useful but dreary work of laying transatlantic cables.

Midway in the 19th Century, a French consular official, Ferdinand de Lesseps, envisioned another route to the Orient—a canal across the Isthmus of Suez in northeastern Africa. And once de Lesseps' dream materialized in the Suez Canal, the advantages planned for the *Great Eastern* were to be lost, for she would be too large to go through this man-made passage, and the long voyage around the Cape of Good Hope would become too costly once the shorter route was opened.

The idea of such a waterway in the Middle East was ancient; a canal joining the Nile and the Red Sea had, in fact, been cut by Egyptian kings in the Seventh Century B.C., and had remained intermittently in operation for more than 1,000 years. But until de Lesseps took up the task, a short route to the East was only a memory and a dream.

De Lesseps was an idealistic entrepreneur whose motive for building the Suez Canal was expressed in a motto: *Aperire terram gentibus*—"To open the world to all people." De Lesseps offered shares in the venture to all Western powers, expecting that they would leap at the opportunity. But not all 19th Century entrepreneurs shared de Lesseps' idealism or his vision. The British government, in command of all existing sea lanes and of most Eastern trade, suspected that the Canal was a French plot to colonize Egypt, to threaten British trade and to interfere with Britain's rule in India; when offered 80,000 shares in the venture, or one fifth of the total, the British declined to buy any at all.

Officials of the still-youthful United States government, wary of international entanglements, refused to buy the 20,000 shares de Lesseps offered them.

In the end the Egyptian government took more than 177,000 shares, for a seven-sixteenths interest in the venture; of the remainder, some 207,000 shares—over half the total—went to 21,000 Frenchmen, most of them small investors. Only 188 persons bought holdings larger than 100 shares; the average holding was nine shares. Here was modern capitalism at work: in a project of worldwide significance, tens of thousands of modest individuals had a financial stake.

After 10 years of construction the Canal was opened on November 17, 1869, to the biggest international fanfare since the Great Exhibition. At dawn of that day more than 80 men-of-war, merchant vessels and other craft from all the countries of Europe rode at anchor in the harbor of Port Said. Shortly after 8 a.m. a cannon boomed, and the French imperial yacht, l'Aigle, with the Empress Eugénie on the bridge, lifted anchor and led the flotilla into the new Canal. Royal ships from the other nations followed her at 10-minute intervals. A frigate carried the Emperor Franz Joseph of Austria, then came a gunboat with the Crown Prince of Prussia on board; the Prince and Princess of Holland were next. Queen Victoria's Ambassador to Constantinople stood at the railing of a British Admiralty yacht. From the shore the cheers of Turkish nobles and Arab workmen, Egyptian soldiers and Greek sailors, Syrian merchants and Sudanese warriors vied with wailing sirens, shrill steam whistles and blaring military bands.

More than any other single project of the 19th Century, the Suez Canal revolutionized world trade. Cutting through a hundred miles of desert and eliminating the voyage around the Cape of Good Hope, it reduced the distance from England to India by more than 4,000 miles. In 1870, its first full year of operation, 437,000 tons of shipping passed through it. Two thirds of that shipping was British. Once the Canal was in use, therefore, the British had second thoughts about owning a share in it, and political conditions in Egypt gave them another chance.

The Khedive of Egypt was a profligate ruler who nearly bankrupted his country. In 1875, finding himself heavily in debt, he offered to sell his shares in the Canal. British Prime Minister Benjamin Disraeli moved swiftly to edge out the French, who also had an eye on the shares; he quickly raised four million pounds and acquired Egypt's seven-sixteenths control for the British government.

The Suez Canal became the busiest of the world's waterways. Within a few years of its opening, a worldwide business empire, run from Europe, was firmly established. The complexity of the transactions involved in the growing world trade had in the meantime given rise to new developments in finance. This was the era in which modern capitalism took firm root.

Capitalism per se was not new; it had begun to develop with medieval commerce. The advance of the Industrial Revolution, with its ever-more-expensive machinery, had increased the use of partnerships, in which two or more entrepreneurs joined together, as James Watt and Thomas Boulton did to manufacture the steam engine; one supplied the mechanical genius and the other the money to put him in business. By the middle of the 19th Century, bankers and financiers had begun to sell small shares in large enterprises to a great many investors, thus distributing ownership and separating it from management.

This system was accelerated by the needs of the early railroad corporations, which demanded such vast amounts of money to begin operation that perforce they enlisted large numbers of investors. The system was soon adapted to the new steamship lines, canals and other substantial undertakings.

With the distribution of ownership came the principle of limited liability. Under the old partnership and its bigger relative, the joint stock company, an investor was liable to the full extent of his private assets for the debts incurred by the company in which he had invested; he could lose his entire estate if the company went bankrupt. In the new limited-liability company, an investor lost only the amount of his investment if the corporation failed. An added advantage was that the investor could spread his capital among many enterprises rather than risk all he had in one.

The new kind of company won ready acceptance, in government circles as well as in business. In the 1850s European nations began to sell limited-liability stocks and bonds in order to finance state projects—the building of harbors and other public works, the purchase of armaments and even the waging of war.

As financial transactions spread out over the world, the investment bankers who handled them grew in power. Because of the vastness of the sums involved, many such bankers became involved with the policies of the corporations and the governments with which they negotiated. Some of them even rivaled in fame the sovereigns they served. Of these, none were more spectacular than the Rothschilds.

What other investment bankers did on a large scale, the Rothschilds did on a magnificent scale. By the middle of the 19th Century their family constituted a dynasty. It had been founded in the Jewish ghetto of Frankfurt am Main by a small money-changer, Mayer Amschel Rothschild, at the end of the 18th Century. Mayer's five sons carried on his business; indeed, they turned it into an empire which together they ruled from Frankfurt, Vienna, London, Paris and Naples. There were few major European financial enterprises, public or private—from railroad stocks to government bonds—with which one or another member of this incredible family was not associated.

The Rothschilds did their business with great aplomb and effectiveness, and in pursuit of a project they could overcome any opposition. In the 1830s, for example, plans for the first French railway aroused a storm of shrill Gallic indignation: fire from the engines would burn crops and forests, the roaring engines would send people fleeing from the countryside and cause cattle to stampede. James Rothschild, a son of Mayer's who was established in Paris, calmly ignored the public clamor and persuaded the government to sanction, first a railroad from Paris to Saint-Germain, 11 miles away, and then another from Paris to Versailles. Having successfully completed both of these, Rothschild planned a third, the Chemin de Fer du Nord—a gigantic project that would connect Paris with the industrial north.

Again there was protest, but again James secured a government license. This time he floated a stock issue of 150 million francs and coolly gave seven and a half million francs' worth of the shares to government ministers and to journalists. Suddenly the project had so much government and press support that Frenchmen flocked to buy shares. When the Chemin de Fer du Nord was ceremoniously opened in 1846, James accepted the congratulations of the Court, the legislature and the press on the completion of "his" railway. He had profited, to be sure, but his Chemin de Fer du Nord was to prove an immensely important national asset, for it pushed France a good way forward in the Industrial Revolution.

The Rothschilds exercised many of the prerogatives of royalty. They maintained their own corps of private couriers whose blue and yellow caps were recognizable throughout Europe. They also traveled among royalty and extended hospitality to

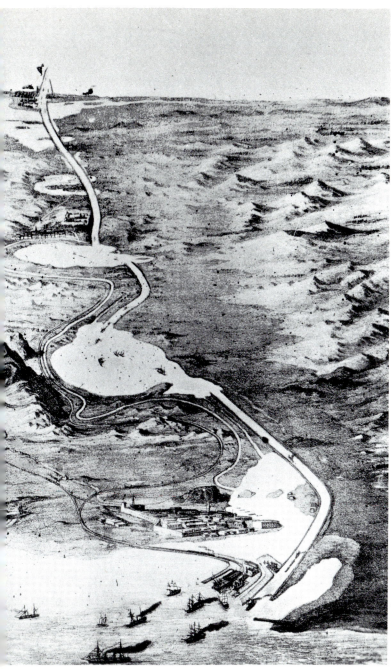

LINKING EAST AND WEST, *the Suez Canal provided Europe with an easier and shorter route to the treasures of the Orient. The canal, as shown in this contemporary drawing, ran about 100 miles from Port Said on the Mediterranean to the Gulf of Suez in the Red Sea; it was 190 feet wide and 26 feet deep.*

crowned heads. In February of 1862 the Emperor of the French, Napoleon III, paid a visit to James Rothschild, who had just moved to a magnificent new estate at Ferrières. Rothschild greeted Napoleon at the entrance to the château, then led him through salons adorned with paintings by Van Dyck, Rubens, Giorgione and Velasquez. At luncheon Rothschild and his royal guest dined from Sèvres porcelain decorated by Boucher and listened to music composed for the occasion by Rossini. That afternoon, in the great park of Ferrières, a hunt was arranged at which more than a thousand head of game were shot. It had been the visit of one emperor to another.

Some years later the King of Prussia, using this resplendent château as temporary headquarters during the Franco-Prussian War, is said to have exclaimed: "Folk like us can't rise to this; only a Rothschild can achieve it!"

In his financial dealings James Rothschild was as imperial as his sovereign and equally able to do as he pleased. He could decline to float the stocks for so lucrative a project as the Suez Canal when De Lesseps objected to paying the Rothschild 5-per-cent commission. A few years later James's son Alphonse, by rallying all the important bankers of Europe, raised the five-billion-franc indemnity imposed on France by Germany after the disastrous Franco-Prussian War of 1870-1871. Behind Alphonse stood cousins in all the capitals of Europe, so rich and so well-connected that they could undertake among them to meet a war debt that staggered a nation.

In London, James's nephew Lionel raised eight million pounds for the British government for the relief of the Irish famine in 1847. (This was no act of philanthropy; for aiding the starving, Lionel drew the family's usual commission.) In 1854 he floated a 16-million-pound bond issue in order to finance the Crimean War. And when Disraeli decided to buy an interest in the Suez Canal in 1875, he went to Lionel Rothschild—the one banker in England who could provide the necessary £4,000,-000 on short notice.

The Rothschilds, with their wealth, their power, and their hand in royal and state affairs, illustrate the culmination of a social as well as an economic revolution. The rise of medieval commerce had first brought into being a bourgeoisie, or middle class. Now in the 19th Century capitalism thrust the bourgeoisie forward and expanded its numbers greatly.

At the pinnacle of the new social structure, in a position of aristocracy that had little to do with lineage, and rivaling in power the old titled nobility, reigned a small group of capitalists whose monumental wealth came from investments in the labors of others. Under them ranked a sizable and many-complexioned "middle class," men engaged in the professions, in manufacture, in commerce, in civil service or in shopkeeping; they worked for their living but earned sufficient surplus to participate in the capitalistic adventure. This group, together with the wealthy financiers, were primarily interested in the pursuit of profit unshackled by government interference—a philosophy that would ultimately be called free enterprise.

At the bottom of the social structure, and separated from the middle class by a severe and widening cleavage, stood a large and newly created proletariat—artisans, miners, skilled mechanics and unskilled laborers, most of them uprooted peasants and craftsmen, who struggled to survive on precarious wages, and who had neither share in ownership nor surplus earnings to spend for investment or amusement.

Free enterprise was chiefly a middle-class preoccupation, but not exclusively; all monarchs and their states benefited from the commercial expansion of their subjects, and one prince of the blood went into business himself. That was King Leopold II of the Belgians; of all the exponents of free enterprise during the 19th Century, none was more enthusiastic than he. In 1876, aware that interior Africa was being penetrated by white men, Leopold formed the International Association for the Exploration and Civilization of Africa. A year later, hearing that the English explorer and journalist H. M. Stanley had traveled the length of the Congo River to its mouth, Leopold began to think of the commercial possibilities of the region. When Stanley returned to Europe in 1878, Leopold summoned him to Brussels, then commissioned him to survey the river basin, build a chain of forts and trading posts and make agreements with the native chiefs.

Within five years the Belgian King had carved out an empire of 900,000 square miles. Leopold solemnly proclaimed that "to open to civilization the only part of our globe where it has not yet penetrated, to pierce the darkness which envelops whole populations, is a crusade, if I may say so, a crusade worthy of the century of progress."

Leopold did not acquire the Congo as a Belgian state possession, but as a fief of his own. In 1891, when the Congo's wealth in ivory and rubber was revealed, Leopold established a monopoly on these treasures—and made a personal fortune of $20 million. In 1908, after it was disclosed that this fortune had been extorted from the natives by cruel treatment, the angry Belgian government took over the colony and made it a state possession.

By the time Belgium took over the Congo from Leopold, most of the major European states had followed their subjects into overseas territories to settle and administer them. They all rationalized, as Leopold did, that they were bringing light into the darkness—and many of them were. If they exploited resources and labor—even as they spoke of the White Man's Burden, of *la mission civilisatrice*, of the spread of *Kultur* or of Manifest Destiny—

they also helped to spread material progress throughout the world. They built great dams and aqueducts that brought modern systems of sanitation and irrigation to Egypt and India. They curbed tribal warfare, improved communications and transportation, and cut down disease. They fed the people and taught many of them to read and write. They also put a stop to long-established slave trading.

Most of the undertakings, to be sure, were initiated solely for commercial profit—a motive so compelling, and so universally accepted as the norm, that not for a long time did it come into conflict with national loyalties. In the Crimean War during the 1850s, London banks had floated loans for their enemy, the Russian government, and no one thought any less of the bankers for it; business and politics were supposed to be independent of each other. But less than half a century later, when Germany in concert with other Western powers was suppressing the Boxer Rebellion in China, the munitions manufacturer Fritz Krupp drew a sharp reprimand from Kaiser Wilhelm II for selling guns to the Chinese. "This is no time," the Kaiser angrily wired his subject, "when I am sending my soldiers to battle against the yellow beasts, to try to make money out of so serious a situation." By the end of the century, free enterprise, even as it fanned out internationally, had inextricably woven economics into the fabric of national politics.

Two governmental policies underlay the economic expansion of the 19th Century. One was Free Trade, the absence of tariff barriers to export or import. The other was the gold standard, the valuation of currency in terms of a given weight of gold.

Free Trade—a reversal of protectionism that dated back to the Middle Ages—was adopted first by England, in the 1840s, when it saw a lucrative market abroad and had no competition at home from foreign manufacturers. Other countries, believing that Free Trade was the magic formula by which England had put itself so far ahead in industry and commerce, followed its lead and lowered tariff barriers in the 1850s and '60s.

But political actions usually result in reactions, and before long there was a widespread reversal of Free Trade. In the 1870s, after repeated and persistent business fluctuations, Continental manufacturers became painfully aware that in depression periods they could not meet the competition of inexpensive foreign products on their own markets. Industrialists therefore ceased to regard Free Trade as the source of prosperity and began to view it instead as a source of trouble.

As a result, Continental nations began a swing toward protectionism. The movement was initiated in the 1870s by manufacturers in the partially industrialized nations (Russia, Spain and Italy) and it was then taken up by Germany and France. Soon the land-owning farmers of these nations joined the industrialists in clamoring for government protection against foreign competition. Farmers had supported Free Trade so long as they produced a surplus for export and could buy foreign farm machinery. But when Russia and the United States began selling significant quantities of grain on European markets, domestic agriculture was threatened.

The cries of the farmers fell on sympathetic ears in government circles. Fearing that diminishing food production, if not checked, would render them vulnerable in time of war, European governments espoused Alexander Hamilton's theory that "Every nation ought to endeavor to possess within itself all the essentials of national supply." Partly on this rationale, and partly out of a desire to raise revenue for more public works, for social reforms, and for manufacturing armaments, most of Europe reinstated the tariff in the 1880s. Only Great Britain, Belgium and Holland resisted the swing toward protectionism; in those countries commercial interests outweighed agrarian concerns.

Even more advantageous to prosperous trade than the absence of tariffs was the international adoption of the gold standard. Britain had adopted that standard in 1816, when it defined the pound sterling as the equivalent of 113 grains of gold. Countries that were later in entering the Industrial Revolution maintained confusing and archaic systems of finance until past the middle of the century; the German states, for example, used two different systems of silver coinage and a combination of 140 kinds of paper currency. But hand-in-hand with increased industrialization came monetary reforms and then, when industrial nations were rich enough to buy it, came the discovery of new deposits of gold in California, Australia, South Africa and Alaska. By the end of the 1870s most of the industrialized nations had accumulated gold reserves and had defined their currency in that precious metal, thus making possible orderly settlement of international accounts. A person holding any standard currency—pounds, francs, marks, dollars—could turn it into gold or into another currency on demand.

The sweep of prosperity was generally upward, but it was not unbroken, and the 19th Century experienced a new phenomenon—the world economic cycle. Earlier eras had known business crises and depressions, but these had been ascribed to war and acts of God, such as famine and plague. With the rise of modern capitalism came periods when crises were no longer attributable to catastrophes of nature, but arose out of production and its relation to financial structure. Sometimes manufacturers expanding in search of greater profit found their plants swollen to overcapacity and therefore laid off workers. At other times the rich engaged in reckless speculation; then followed collapses of the stock market, runs on banks, bankruptcies, shutdowns and unemployment. As industrial capitalism spread, financial crises were no longer confined to single areas, as agrarian crises had been, but were felt all over the world. Moreover, they recurred with more frequency.

The first collapse solely attributable to capitalism gave a dramatic demonstration that the new economy constituted an interdependent world network. In August 1857 a Cincinnati insurance company failed after the head of its New York branch either helped himself to the company's funds or lent them rashly to a railroad in which he was interested (it is not clear which). As it happened, the firm served as a clearing house for several banks, railroads and other corporations across the United States. The insurance company found itself short of money and unable to meet its obligations—and every firm in which it had an interest was shaken to the core. Banks closed, railroads went bankrupt; soon almost 5,000 businesses had collapsed, and the prices of raw materials and agricultural products fell simultaneously. Within three months the effects had spread to northern England, Scotland, Hamburg and Paris, and ultimately to Moscow, Odessa and the coastal cities of Latin America. So intricate was the capitalist economy now that one man's mischief sent reverberations around the world.

Despite intermittent financial crises, the great saga of European acquisition was to last until 1914, and it was for the most part a magnificent and cooperative enterprise. But it was a period marked by contrasts. Great wealth did not wipe out poverty; national achievement was breeding international animosities; violent economic fluctuations were making it apparent that the new economic order was bringing insecurity along with profits, losses along with wealth. If the extension of the capitalistic spirit gave riches and purpose to many, it brought insecurity and despair to others—a development that was to have grave effects on society. The prevailing spirit of the age was the worship of material and commercial gain; nevertheless, some voices began to cry for social reform.

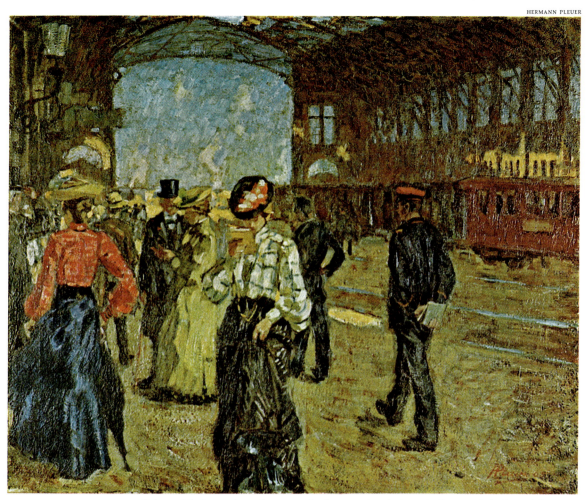

WAITING FOR THE TRAIN, *well-dressed ladies and gentlemen restlessly pace the platform in Stuttgart's high-roofed station.*

A HOLIDAY BY RAIL

The mid-19th Century was a time of great excitement, as railroads reached out their steel tentacles all over Europe. In the wake of the railroad came new benefits in prosperity, leisure and travel. Just riding the trains at a breathless 55 miles an hour gave life a whole new dimension of speed. To be sure, the third-class passenger rode on hard benches, froze in winter, and gulped his meals on station platforms while the train waited (or departed). The wealthy, on the other hand, luxuriated with velvet cushions, foot warmers and champagne in the new dining cars. But for everyone with the price of a ticket the railroads meant the freedom to see and do things they never had before.

One of Europe's chief attractions was Paris, with its bright lights and broad boulevards, its tranquil countryside and, not far distant, that new middle-class institution—the seaside. The French Impressionists, then at their height, caught the spirit of such holiday spots in brilliant canvases that immortalized the age.

SPANNING THE SEINE, *a train crosses the bridge at Argenteuil, a suburb north of Paris.*

THE SOOTY ELEGANCE
OF TERMINALS AND TRAINS

By the late 1800s France could boast one of the most modern railroad systems in all Europe: approximately 11,000 miles of track had been laid, and the dashing French trains were celebrated as models of efficiency. Unfortunately, speed had to be curtailed to 55 miles an hour following several mishaps. But gradually safety features were added: platforms were raised after several passengers toppled off the train steps and rolled under the wheels, and new type "T" rails replaced the old strap iron rails that sometimes worked loose to derail trains or to spring up and pierce unlucky passengers.

Along with railroad efficiency France excelled in bringing style and elegance to railroad architecture, a passion of the day. Before 1850 rail terminals were often crude, but after mid-century they began to outdo each other in size and splendor. Some were compared to medieval cathedrals; others took inspiration from the Pantheon and other stately public buildings. Attached hotels were equally magnificent: the one beside the Gare St. Lazare was frosted with mansard roofs and chimneys copied from the Tuileries. But guests often complained that it was so noisy they couldn't get a decent night's sleep, and no one had yet thought of how to eliminate the soot, smoke, cinders and noise.

A MAJESTIC TERMINAL, *the Gare St. Lazare was one o*

seven stations that welcomed rail travelers to Paris. Above the smoke, bustle and snaking tracks of its shed soared a simple, stunning roof of iron and glass.

PROMENADING *was a favorite pastime in 19th Century Paris. The* *elegant Boulevard des Capucines (above), with its carriages and* *cafés, drew visitors who liked to mingle with the smart crowds.*

BUYING A HAT *(right) was a temptation few women visitors to* *Paris could resist. And such hats! They came decorated with fruit,* *flowers, birds, feathers, plumes, bows, lace and long ribbons.*

PLEASURES OF PARIS

BROWSING *in the book and art stalls could turn up treasures; this gentleman leafs through a folder of prints.*

HONORÉ DAUMIER

Paris and its sparkling reputation drew visitors from all over Europe. It was acclaimed the most beautiful city in the world, with its noble architecture, elegant women, smart shops, celebrated fashions and racy night life. Visitors strolled the boulevards and savored the city's many pleasures—the parks and monuments, libraries, museums, the soaring Eiffel Tower, the renowned Champs-Elysées. The very air seemed different; a character in a Maupassant story compared it to "taking a bottle of champagne." The middle-class visitor breathed deeply of this rarefied air; whether he took his champagne from the atmosphere or out of the bottle, Paris could satisfy the most exotic thirst.

EDGAR DEGAS

EDGAR DEGAS

WATCHING A STAGE SHOW *was a high point of the evening in the city's livelier cabarets, like Les Ambassadeurs (left). Other popular night spots were The Dead Rat, The Casino of the Concierges, and The Bride's Bedtime.*

SIPPING AN APERITIF *in Montmartre's famous Moulin Rouge (right) was something to talk about on returning home. Here, as elsewhere, champagne flowed, and there was dancing, laughter and love-making in private alcoves.*

PIERRE BONNARD

A NIGHT ON THE TOWN

Emperor Napoleon grandly declared his reign a time of peace and prosperity, and it was then that Paris won her enduring reputation for gaiety. By day the city was industrious, even sober, but at dusk a special magic descended. The cafés began to fill and there was the hum of good conversation above the tinkling of glasses.

After his apéritif, the new arrival could choose among diversions both serious and frivolous. In one outdoor dance palace, the Closerie des Lilas, the pungent scent of a thousand flowers wafted over couples whirling to such dances as the mazurka, the polka and cancan. In another celebrated dance hall, in a dazzling blaze of light, a 50-piece orchestra played under a canopy of fake palm trees with dim gas lamps suspended among their waxy leaves. Hooped crinoline skirts billowed to show expanses of lacy petticoats, and necklines dipped perilously low. Everyone drank champagne and laughed at songs like "It Tickles My Nose."

65

OUTINGS IN THE COUNTRY

The burgeoning railroad system not only made long-distance travel possible; it also brought vast changes to the suburbs. These environs were served by local trains which shook sleepy little rural communities awake and linked their fortunes to the cities. The trains left Paris every hour; they had an odd

CLAUDE MONET

STROLLING THROUGH OPEN FIELDS *was a popular diversion for Parisians, made possible by the trains that served the suburbs.*

EDGAR DEGAS

ATTENDING THE RACES, *two lady spectators watch from a carriage as jockeys warm up their mounts.*

arrangement of two floors, the upper one with windows, the lower one a primitive box with just slits in the sides for those who insisted on fresh air. For city people these local trains meant a new mobility—visits to the homes of friends in the suburbs, picnics, horse races, swimming in the Seine.

At the same time a new pattern of life was emerging: ugly little industrial towns began to erupt along railroad rights of way, and the whole tempo of life speeded up. The suburbs also began to bloom as "bedrooms" for the city, as many of the wealthy moved there, soon followed by the middle classes.

ENJOYING THE RIVER, *Parisians bathe and dine at La Grenouillère,*
a recreation spot on a branch of the Seine just outside the city.

GAY WEEKENDS BY THE SEA

As the railroads reached the ocean, little seaside resorts sprang up by the dozens all along the coast. No longer were these cooling havens the special province of the aristocracy; now they could be reached by everyone. The resorts were served by trains that left Paris after work on Saturday and returned early Monday morning. With unconscious irony, they were called "pleasure trains"; the fact is, they involved many hours of hard travel for a few refreshing moments on the beach.

The resort towns, in turn, became known as "pleasure towns," and many grew famous through-out Europe for their elegance. Vacationers stayed in all kinds of hotels, often magnificent places with 200 or more rooms, an exquisite birdcage elevator, inviting terraces and a dining room with large plate-glass windows that afforded a magnificent view of the sea and promenade. The wealthy sometimes rented private villas for the whole summer, but few people could afford extended stays. Most middle-class visitors had to get back to the city after the weekend; no frugal merchant would think of closing his business during the summer, and the custom of taking long vacations had not begun.

CAMILLE PISSARRO

A RESORT TRAIN *pulls into the busy port of Dieppe on France's northern coast while throngs of gaily attired vacationers wait beside the tracks. Each train arrival was a major social event.*

A RESORT HOTEL, *like this one at Trouville, offered guests many pleasures: strolling on the boardwalk, sipping drinks under an arbor, or just sitting back and staring out at the sea.*

CLAUDE MONET

THE FULL-DRESS JOYS
OF BEACH LIFE

Most guests at a seaside resort spent the day on the beach swimming or relaxing, or perhaps taking the waters at a nearby health spa. Many ventured on the sand only for the bracing air, the ladies often shielding themselves from the sun with a dainty parasol *(above, left);* a suntan was considered inelegant. One of the more delicate problems of beach life was where to change one's clothes. This led to the creation

of so-called "dressing machines," little boxlike houses on wheels that could be rolled next to the surf. A lady would chastely enter and change, but usually she came out more dressed up than she went in: bathing suits were complex ensembles consisting of billowing bloomers and elaborate tunics, caps or hats to protect the head, and high shoes—not even ankles were exposed. In the evening vacationers would meet in the cafés, gather in the casino to play *chemin de fer* or promenade along the avenues beneath graceful Normandy poplars and spreading lindens. Then, suddenly the weekend was over, and ahead was the long, exhausting train ride back to the city, and work. But the homebound vacationer at least had one cheering thought: next Saturday, if all went well, he could return for another exhilarating weekend by the sea.

4

THE WORKERS' PLIGHT

One midnight in the 1860s a reporter for the New York *World* stood on London Bridge and looked out over the port of the richest city in Europe. "The Thames lay at our feet," wrote Daniel Kirwan, "spread out like a map. . . . Below us, to the left, the Catherine Docks, full of shipping; the London Docks, full of shipping; Shadwell lined with lighter craft . . . millions of masts. . . ." But underneath the bridge, in the shelter of its arches, Kirwan came upon a different scene, "a perfect gypsy encampment" of human wreckage. "Eight of these persons were of the male sex, and besides these there were two old haggard-looking women and a grown girl of 20 years or thereabouts, and a child of 10 years, in all the glory of rags and destitution."

The contrast between this wretched group and the prosperity flaunted by the millions of masts did not escape Kirwan. Poverty was nothing new, but poverty in a society capable of producing enormous wealth was difficult to ignore. In fact, it was the great riddle of the Age of Progress. The inadequate living standards of great masses of people mocked the marvelous advances of science and industry, and troubled the social waters of Europe throughout the 19th Century. Among other things it contributed to the brief and bloody 1871 uprising called the Paris Commune and later to the Russian Revolutions of 1905 and 1917. The Industrial Revolution, which had brought men new freedom, new opportunities, and new comforts and conveniences also had brought them a new problem: a huge urban working class without any of the traditional safeguards to economic security.

Before the Industrial Revolution society had been composed principally of two classes, aristocrats and peasants. The former owned the land and the latter lived on it in servitude, tilling the soil and caring for the livestock, generation after generation. The system had its evils but it offered the peasant certain rights in return: he could look to his landlord for help and support in time of trouble, and his place was secure in the scheme of things. The Industrial Revolution threw this feudal world into chaos. Mechanized farming disrupted age-old agricultural patterns, and the abolition of serfdom released the peasant from his bondage to

LEFT BEHIND BY PROGRESS, *a woman sits on a London doorsill with teakettle and cup beside her (left); she epitomized the paradox that, while the Industrial Revolution created great wealth, it also created misery and poverty.*

the land. Feudalism was gone from England and France by the end of the 18th Century, from central Europe by the middle of the 19th. By 1861 it largely disappeared from the vast territories of the Russian Empire when Czar Alexander II freed more than 20 million Russian serfs.

Unable to make a living on the land, and deprived of his traditional income from craftwork, the peasant sought work in industry. Factories and mines offered a living, however precarious, plus the attraction of being paid in wages rather than in kind. Peasants migrated in great numbers to the new industrial centers of England and France, Belgium and Germany, Austria, northern Italy and Russia. They provided the labor for the great new textile mills of Lancashire and Alsace, of Saxony, Bohemia and Tuscany, and for the mines of the English Midlands, Liège and the Ruhr.

For the most part these new industrial workers simply exchanged one kind of bondage for another. The laws which freed them from serfdom also freed them to starve. With no property of their own and no use for their rural skills, they were entirely dependent for their survival upon the capitalist who owned the factory or mine, who controlled what Marx was later to call "the means of production." The capitalist owner could hire or fire at will, and he drove the hardest bargain he could for wages. When he needed workers, he paid well, but when goods were plentiful he paid what he chose—sometimes less than a living wage. He could even lock out his workers whenever he chose, and did so without pity. Between the insecurity of a boom-and-bust economy and the inhumanity of the working conditions, the plight of the early industrial workers was, with few exceptions, appalling.

In Belgium in 1885, the annual report of an inspector general of prisons and charitable institutions disclosed that the amount spent on feeding and clothing one convict exceeded the yearly wages of a whole family of workers—father, mother, children. In 1860, a county magistrate in Britain reported that children employed in the Nottingham lace trade were "dragged from their squalid beds at two, three or four o'clock in the morning and compelled to work for bare subsistence until 10, 11 or 12 at night, their limbs wearing away, their frames dwindling, their faces whitening, and their humanity absolutely sinking into a stone-like torpor, utterly horrible to contemplate."

One small boy employed in a British pottery mill in 1863 supplied this account of his work to a government commission investigating child labor: "I turn jigger, and run molds. I come at 6. Sometimes I come at 4. I worked all last night, till 6 o'clock this morning. I have not been in bed since the night before last. There were eight or nine other boys working last night. All but one have come this morning. I get three shillings and sixpence. I do not get any more for working at night." Again and again, in other accounts, the dreadful litany is repeated.

To add to the horrors of industrial conditions, cities in which the factories were located grew suddenly and enormously, without plan. Between 1800 and 1900 the population of Europe more than doubled—there were fewer wars to kill people off, and science was improving man's chances of staying alive. Most of this huge new population earned its living in industrial towns. By 1851 there were more Englishmen and Welshmen living in cities than in the countryside, and by 1901 the proportions were three to one. London, with a population of 4.5 million, held one tenth of the entire population of England and Wales. Elsewhere the story was the same; Paris was home to one out of every 25 Frenchmen; one of every 20 Germans lived in Berlin.

Migration was also a mark of the times, and this, too, contributed to urban confusion. The poor and oppressed were constantly on the move in search of work and a better life—from country to town, from city to city, across frontiers and oceans. The Irish fled Ireland by the thousands during the great potato famines of the 1840s; the Jews fled Poland in the 1880s to escape the anti-Semitic pogroms. Between 1871 and 1900 some 25 million people left Europe for America, Australia and other parts of the world. The mass movement of peoples in the 19th Century, without precedent in human history, was halted only by the outbreak of the First World War.

During the early decades of the Industrial Revolution, living conditions in the cities were terrible. Families were crowded together in jerry-built tenements or in the cast-off housing of the rich. They lived in attics and cellars, sometimes in only one room, sometimes in the corner of a room. In one house in Spitalfields, a London slum, 63 people lived in nine rooms, each of which had only one bed. New housing built especially for the workers was often, in the words of one Royal Commission, "of the commonest materials, and with the worst workmanship . . . altogether unfit for people to live in." Walls were sometimes only half a brick thick, and the builder skimped on windows to avoid the government's revenue-raising window tax. Drainage was wholly inadequate and so were the sanitary facilities. As often as not, sewers were open trenches running down the middle of the street. Pure water was a luxury, and parks and trees in working-class areas were nonexistent.

In these dark, foul-smelling warrens of humanity, simply keeping alive was a major problem. Great epidemics of cholera, typhoid and typhus raged through the cities periodically until the last quarter of the century. The "white plague" of tuberculosis was commonplace, and so were diphtheria and the many diseases associated with malnutrition. "The imagination," notes the English scholar G. M. Young, "can hardly apprehend the horror in

which thousands of families a hundred years ago were born, dragged out their ghastly lives, and died: the drinking water brown with faecal particles; the corpses kept unburied for a fortnight in a festering London August; mortified limbs quivering with maggots; courts where not a weed would grow, and sleeping-dens afloat with sewage."

These conditions degraded the human spirit. Working-class parents thought nothing of drugging their babies with elixirs of opium to keep them quiet, or of putting small children out on the streets to fend for themselves, prey to every sort of corruption. "Vice is in every glance of their eyes," wrote Daniel Kirwan of such "latchkey" children in London's Lambeth quarter. "Crime has already made its graven lines on their young faces." And in Paris the Goncourt brothers, roaming the poorer sections of the city in search of material for their realistic novels, came upon "beggarly children, laughing ferociously" as they watched a young man being attacked and beaten to the ground. Meanwhile, a crowd "like a circus audience . . . feasted its eyes upon this butchery without the slightest evidence of revulsion."

The conditions that assaulted the senses of visitors to these working-class neighborhoods emphasized the widening gulf between upper and lower classes. The two had always lived in different worlds; now the differences were extreme. There were those who had, and those who had nothing— not even an equal chance to stay alive. A child born to an upper-class family in Liverpool or Manchester or Leeds might hope to reach the age of 38, but the average life expectancy of a working-class child in one of those cities was little more than 17. The death rate in London's fashionable West End was half that of the East End slums. Society was split in two; capital and labor faced each other across an impassable chasm. It was this chasm that provided Karl Marx with his theme and prompted

Benjamin Disraeli to observe that Europe had "two nations"—the Rich and the Poor. "We must prepare for the coming hour," Disraeli wrote. "The claims of the future are represented by suffering millions."

Society was split in other ways too. In the old days of village life the landowner and his peasants saw each other daily, but the rich capitalist had very little contact with the workers. He lived in a different part of town, often miles distant, and met the worker infrequently even in the place of work. Few mine or mill owners descended into the coal pits or appeared on the factory floor with any regularity, and most of them assumed that their workers' lives away from the job were the workers' own responsibility. Many rich capitalists, in fact, were only indirectly involved with the workers in any way. The source of much of the wealth during the heyday of the Industrial Revolution, especially in England and France, was stocks and bonds. As they gathered in their dividends and profits, the investors did not inquire too deeply into the conditions that prevailed in the mills that produced them.

This detachment went hand in hand with the 19th Century capitalist's allegiance to *laissez-faire* and Free Trade. Having finally gained the right to operate without government control, entrepreneurs were loath to give up any part of that freedom. When there were fluctuations in the economy, they argued that such events were unavoidable, the result of "Natural Laws." Thus there was a Natural Law of Supply and Demand, a Natural Law of Diminishing Returns, an Iron Law of Wages, which made labor a commodity like any other commodity, subject to the fluctuations of supply and demand. The owners of capital believed that these laws were immutable and that if they were permitted to function without restrictions they would eventually lead to the greatest

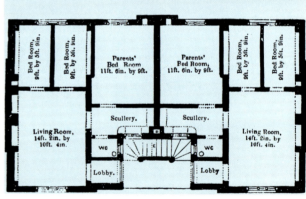

A PRINCE'S PLAN, *these low-cost apartments for working people were designed by Prince Albert. He thought each unit should rent for about three shillings a week. The floor plan specified three bedrooms, a kitchen and parlor. Only one of the four-family brick houses was ever built—a model shown at the Great Exposition.*

good for the greatest number. In practice, they led to great profits for a very few and to great misery for countless others. Mistaking freedom from control for license to do as they pleased, capitalists indulged in practices that took no account at all of the worker's right to a living wage and decent working conditions.

Inevitably this large, and largely neglected working class came to be one of the gravest and most persistent problems of the Age of Progress. It was attacked with equal concern by coldly practical politicians and by warm-hearted humanitarians, by conservative governments and by liberal-minded men of affairs. For some the spur was moral indignation, for others it was the threat of revolution. The solutions they proposed ranged from factory laws to model factory towns, from free public education to extension of the franchise to all men, regardless of their condition in life. There were advocates of public baths and wash-houses for the poor, of visiting hours for museums and libraries that would make them available to workers, of concerts in the parks on Sunday, the workers' day off. Other public-spirited citizens, worried about the health of all classes, began to speak out against impure foods sold in the markets, against dirty bakeries, and against such practices as the addition of alum to bread dough to make it rise and the use of plaster of Paris and poisonous food colorings to make confections more appealing to the eye.

Most of these pressures for reform of all kinds came from above, from within the ruling class. They brought about the changes that might otherwise have been a long time in coming, and relieved some of the worst causes of the workers' desperation. But in the long run the most effective pressures for change came from below, from within the ranks of the workers themselves. Through trade unions and political parties, and through strikes, labor gradually forced capital to come to terms with its demands.

England, the home of free enterprise and the original home of the Industrial Revolution, was the first to feel the effects of the social evils they caused, and the first to attempt to deal with them. Humanitarian reformers like Lord Ashley, one of England's great landowners, sponsored legislation to shorten working hours, to set limits on the age of child labor, and to bar women, girls and very young boys from working in the mines. Every proposal was preceded by an investigation, producing a continual round of Royal Commissions and Select Committees, Blue Books and White Papers. But the resistance was stubborn. It was not until 1850 that a more or less effective Factory Act limited the working day of women and children in all industries to 10 and one half hours.

In this fight there were no clear-cut sides. Conservative Tories found themselves defending labor, while liberal Whigs were often aligned with the most intransigent capitalists. It was, the political diarist C. C. F. Greville wrote, a most "curious political state of things, such intermingling of parties, such confusion of opposition. . . . The Government . . . have been abandoned by nearly half their supporters. . . . The Opposition were divided. . . . It has been a very queer affair."

A similar confusion reigned among factory owners. There were many who fitted the popular image of the arch-reactionary, the man with "thin lips and hard eyes," but others were actually in the vanguard of improvement. In 1850, for instance, Sir Titus Salt moved his textile mill to the country and built his workers a model town, Saltaire. Its 800 houses had kitchens and parlors, separate bedrooms for parents and children, and backyards. There was also a school, a church and a park. And in 1888, W. H. Lever, the founder of Lever Brothers, built for his workers Port Sunlight, a factory town which could boast public gardens and houses with indoor plumbing.

The most celebrated of the reforming factory owners, however, was Robert Owen, whose New Lanark cotton mills in Scotland were models of enlightenment. Owen, who had risen from the working class to become part-owner of the mills, pioneered in the field of labor relations. His measures against long hours, child labor and the practice of fining workers for infractions of factory rules were the basis for many of Britain's labor laws, and his provisions for educating his workers' children helped to set the standards later adopted for public education. Owen was also interested in the idea of the cooperative community, to be owned and run by its inhabitants. He spent four-fifths of his own fortune to establish such a community in America, the village of New Harmony, Indiana, but for reasons all too human the noble experiment failed—some of its citizens were too lazy to work, and some too selfish to share.

The workingman's own agency of reform was the trade union. For a large part of the 19th Century most union activities were outside the law; before 1824, unions were forbidden entirely; and, although they were allowed to exist after that date, many of their activities were considered illegal. Even after the passage of the 1871 Trade Union Act, which recognized unions as the legal representatives of their members, many union activities were against the law. Picketing, for instance, was a criminal offense; in the very year the Act was passed, seven Welsh women were sent to jail simply for shouting "Bah!" at a strike-breaker.

But legal or not, unions grew in size and power all through the 19th Century. Originally they were little more than mutual aid societies, groups of men in a common craft who banded together to provide themselves with insurance against sickness or injury. Gradually, however, they became the means through which labor aired its grievances against management and the social evils of the day. From small local units of potters, cotton spinners, tailors and the like, they expanded into national organizations with extensive funds and articulate officers who spoke compellingly for the workers' demands. The first of the big unions was the Amalgamated Society of Engineers, founded in 1851. Others soon followed. Among them was the large and influential Miners' National Association, whose leader for almost 20 years was Alexander Macdonald, an ex-miner who had worked in the pits as a boy and had attended Glasgow University, where he studied Latin and Greek, logic and mathematics.

The unions lobbied effectively for such legislative reforms as an amendment to the Master and Servant Law, withdrawing the right of employers to jail workers who left their jobs without sufficient notice, and for the passage of the Checkweighmen's Act of 1887, allowing a representative of the miners to supervise the company's tally of each man's take of coal. Most union pressures were applied peacefully, by public discussion and political action, and in negotiation and arbitration with employers. The strike, later a potent weapon of discontented workers, was not a major factor in the British union movement until near the close of the 19th Century, when the London Dock Strike of 1889 was called. Violence was not the way of English workers. Conditioned by England's long history of constitutional reform, they chose to progress slowly and in an orderly fashion. Ironically, although Karl Marx and Friedrich Engels, the century's two most militant socialists, formulated their revolutionary theories from studies of English industrial life, Marxist socialism never made much headway there. The country was, as Engels observed, "the most bourgeois of all nations"—even its aristocrats and workers were middle-class in their attitudes. It remained for another country, coming from behind, to lead the Western world in social reform.

That country was the German Reich, created by Otto von Bismarck in 1871 from 26 German states; the new nation was ruled by a kaiser whose chief minister was Bismarck himself. Almost immediately Germany made swift advances in commerce and industry, and before long it was England's strongest competitor. Like England, Germany faced the problem of social unrest, but Bismarck attacked it head-on and with typical German thoroughness. Declaring that "the state is not only an institution of necessity but also one of welfare," he set in motion a whole series of social reforms designed to make the "propertyless classes" look upon the nation as benign as well as authoritarian. In rapid succession he pushed through the German parliament a series of insurance acts against sickness (1883), accident (1884) and old age (1889). By the first years of the 20th Century, Germany had laws regulating every aspect of industrial life: wages and hours, time off, grievance procedures, safety measures. There was even a law prescribing the location and minimum number of toilets and windows in factories.

The motives behind these measures were not entirely altruistic. Bismarck was a notable advocate of *Realpolitik*—of practical as opposed to idealistic government—and he introduced his reforms out of concern over two dangers: the attraction of socialism for the German working class, and the poor health of the men inducted into the German army. He was attempting to safeguard the German military machine, one of the most important tools of his foreign policy, and to insure the stability of the existing German government. On both scores, he succeeded. No other workers in Europe enjoyed such state-ordained well-being; no others had so little cause to revolt.

Bismarck did not, however, eradicate the social-

ist movement. In Germany, as in the rest of Europe, it continued to attract both intellectuals and the more articulate and informed members of the working class. The goals of socialism were not limited to social reform; Marx and his followers believed that the system could serve as the basis of a wholly new kind of society, for which they had formulated a scientific rationale. Indeed, they opposed social reform that was not accompanied by political reform; they denounced Bismarck's welfare state as pseudo-socialist and pseudo-democratic—or, as Wilhelm Liebknecht, a leading Social Democrat, put it, as "the night-watchman state, the beadle state, which stands as a prison guard over the subject people."

Earlier in the century, well before Marx, other socialists had proposed solutions to the problems of poverty and inequality. Most of them involved the drastic reorganization of society. Disciples of Robert Owen had preached self-governing cooperative communities; disciples of the French nobleman, the Comte de Saint-Simon, had advocated public ownership of the means of production, and a society run by engineers and scientists; disciples of another Frenchman, Charles Fourier, proposed a society divided into small, self-supporting units of about 1,600 people each, with every man doing the work most to his taste. To Karl Marx, the German intellectual who spent most of his life in political exile, these were the schemes of starry-eyed dreamers. Marx believed that the important thing was not to conceive of ideal systems, but to work for the dissolution of the existing one, which was bad and therefore ought to be destroyed. Out of its ashes would arise a new society, whose contours would be determined by the people who had made it. "Philosophers," he wrote, "have only sought to interpret the world in various ways; the real point is to change it."

At the root of Marx's beliefs was a theory about

the manner in which societies came into being. They were formed, he said, not from an idea but from the economic facts of life. Thus the feudal society of the Middle Ages was the product of an agrarian economy in which there were two classes, aristocracy and peasantry, and in which the measure of wealth was land. With the growth of trade, however, a new commercial class arose, bringing with it a new social order. Its members were the merchants and craftsmen—the bourgeoisie—whose measure of wealth was money and whose society came to be called capitalism. Inevitably, said Marx, the bourgeoisie had to clash with the feudal aristocrats because their economic aims were antagonistic. Just as inevitably, he predicted, the working class—the proletariat—would clash with the bourgeoisie, also for economic reasons. This class conflict was what Marx called the "dialectic" of history.

Marx believed that he had discovered a Natural Law. "As Darwin discovered the law of evolution in organic nature, so Marx discovered the law of evolution in human history," said his life-long collaborator, Engels, in his eulogy at Marx's grave. The drive for profit, the goal of capitalism, would eventually lead to the concentration of all wealth in the hands of a very few, said Marx, and to an increasingly large and increasingly impoverished proletariat; in the end this proletariat would revolt and take over the means of production. It was this revolution that Marx encouraged. "The Communists disdain to conceal their views and aims," he announced in the *Communist Manifesto* in 1848. "They openly declare that their ends can be attained only by the forcible overthrow of all existing social conditions. Let the ruling classes tremble at a Communist revolution. The proletarians have nothing to lose but their chains. They have a world to win."

In 1864, at a historic meeting in London, representatives of English, French, German, Polish and Italian workers joined together to organize the First International Working Men's Association. Five years later the first socialist party was formed in Germany. By 1895 there were socialist parties of one kind or another in just about every country in Europe. Even England, the original home of *laissez-faire* liberalism, had its socialistic Fabian Society, two of whose illustrious early members were H. G. Wells and George Bernard Shaw. In France the Chamber of Deputies, the country's lower parliamentary body, had one representative from the Socialist Labor party in 1893; in Italy, the Marxist party elected 12 members to parliament in 1895 and by 1900 had 33. Austria, in the same year, had 14 Socialists in its Chamber of Deputies; Sweden had one. And in Germany, where it had all begun, the Social Democrats were strong enough by 1898 to command three million votes and to control 56 of the 397 seats in the Reichstag; in 1912 this figure had risen to 110.

But Marx's theories did not go unchallenged, and in fact his key condition for changing society—open and armed revolution—did not seriously affect European history until long after his death in 1883. The idea of class conflict had effective opponents. One of the most powerful of them was the brilliant and frail Pope Leo XIII, who in the 25 years of his pontificate almost singlehandedly revitalized the Catholic Church and turned its face toward the problems of the modern world. Leo was an outspoken foe of Marxism, but he was equally critical of capitalism. In a famous encyclical, *Rerum novarum*, issued in 1891, he declared that "class is not naturally hostile to class" and condemned Marxism for making the state more important than the individual man.

At the same time Leo called upon capitalism to change its ways. "Labor," he wrote, "is not a commodity," and it was "shameful to treat men like chattels to make money by." Private property

A MARXIST MEMBERSHIP CARD, *signed by Karl Marx himself, attests that Hermann Jung belonged to the International Working Men's Association, also called the First International. Jung, who was prominent in the labor movement, later became anti-Marxist.*

should be distributed more equally among all people, labor unions should be encouraged and labor's demands should be listened to. Above all, said Leo, the sanctity of the family should be preserved. This could be done only by shortening the working hours, especially of women and children, and paying workers a "living family wage." Unfortunately, these injunctions were more often honored in the breach than the observance, but they did make the Catholic Church officially sympathetic to the demands of labor and they did inspire in most Catholic countries political parties dedicated to what was called Christian Socialism.

Marx's theories were also challenged, perhaps even more effectively, from within the ranks of socialism itself. The English Fabians, for example, were generally opposed to the idea of violence, believing instead that society could be converted to socialism gradually, through due process of law. And in 1889, at the time of the founding of the Second International, the German socialist Eduard Bernstein took issue with Marx's analysis of capitalism's development and its inevitable "impover-

ishment" of society. The real facts, said Bernstein, contradicted this doctrine. Although the rich, as Marx had predicted, were indeed getting richer, the poor were not getting poorer. On the contrary, by 1900 the workingman was able to buy half again as much with his wages as when Marx had formulated his theory some 50 years before. To Bernstein, and to other moderates like him, it seemed obvious that workers with voting rights and political parties of their own could transform capitalism into socialism without revolution.

The truth was that by the close of the 19th Century most of the worst social ills of the preceding generation had been corrected. Workers' grievances were no longer directed at the same conditions that had united their fathers in the fight for social justice. Life was not perfect, but it was getting better. In almost every country—the notable exceptions were Russia and a handful of Balkan states—factory and labor laws had made working conditions more humane and the foundations of social security had been laid. Trade unions were universal and were accepted as bargaining agents for their members. An unprecedented wave of prosperity overflowed into the working class, raising wages to new heights. On every side there were abundant signs that man was finally on the way to conquering the three archenemies of civilization—poverty, ignorance and disease.

These signs were nowhere more evident than in large cities, which were at last becoming healthier places to live in. By 1875 London had completed its great Metropolitan Drainage System and the city of Manchester had built a 96-mile aqueduct to bring pure water from the Lake District. Paris had constructed some 385 miles of underground sewers and freed the Seine from pollution. After France's Louis Pasteur had discovered the role of microbes in disease, England's Joseph Lister discovered how to fight them, enormously affecting the control of epidemics (and earning Lister a peerage from Her Majesty's grateful government). Meanwhile, the work of German scientist-statesman Rudolf Virchow in demonstrating that sewers were breeding places of microbes led to improved public health.

Cities were becoming modernized in other ways as well. First gas and then electricity revolutionized the lighting of homes and streets, while on those same streets motor-driven vehicles revolutionized transportation. After a half-century of unsupervised growth, many cities reintroduced urban planning. Public parks and recreation areas were built and there were even a few attempts at municipal slum clearance. In the 1870s the mayor of Birmingham, Joseph Chamberlain, tore down the shanties and hovels of the poor and replaced them with city-owned housing. Chamberlain also brought the city's gas and water works under municipal control, improving the product and lowering the cost to the public. Two decades later, at the turn of the century, Vienna's mayor, Karl Lueger, municipalized not only the city's public utili-

ties—gas, water and the streetcar system—but even such institutions as orphanages and funeral homes. He also ringed Vienna with a green belt of woods and meadows maintained by the city for public use. In actuality the motive for Lueger's "socialism" was political; Vienna's masses idolized him and re-elected him to office time and time again from the 1890s up to his death in 1910.

Workingmen everywhere had more influence on government. Partly this was direct, through the pressure of the ballot: with the spread of universal male suffrage, the labor vote outstripped all others. Just as often it was indirect, through the pressure of public opinion. The workingman had a newly acquired knowledge of public affairs. Compulsory schooling made him literate, and popular journalism gave him newspapers and periodicals designed to reflect his interests. Between 1868 and 1886 elementary school education became obligatory in England, France, The Netherlands, Belgium, Italy, Germany, Switzerland and Austria-Hungary —in fact, in almost every country of Europe (once again, the chief exception was Russia). By 1900 fewer than 5 per cent of the people of Britain, France, Germany and Scandinavia were illiterate. In the last two decades of the century the number of newspapers in Europe doubled, and the circulations of several of them—*Le Petit Journal* in Paris, the *Lokal-Anzeiger* in Berlin, London's *Daily Mail* and *Daily Express*—catapulted into the millions.

With these developments the workingman became a force in capitalist society. As such he had far less reason to join Marx's international brotherhood of the proletariat. The ties of class became far less important than another sort of fraternity, the common bonds of language and place. It was nationalism, not internationalism, that bound men together in the Age of Progress and gave the era its strongest force for social order: the nation-state.

A CRY OF PROTEST

To many writers and artists observing the sorry lives of industrial workers, the age of "progress" was a grim jest. "Oh God!" wrote Thomas Hood, "that bread should be so dear, and flesh and blood so cheap!" In the late 19th Century, in the Berlin slum where her husband doctored the poor, the artist Käthe Kollwitz also watched gravely, and wielded charcoal and pen so effectively that she became one of the most eloquent voices of protest. Movingly portraying the human by-products of both factory and farm, capturing their sorrows, frustrations and rage *(above)*, she helped in some measure to change the ugly world she saw.

HARNESSED
TO THE PLOW

Workers who remained on Europe's farms were no better off than those who fled to the crowded cities to work in factories. Käthe Kollwitz's dramatic picture of a man harnessed to a plow was more than symbolic; some peasants actually did drag plows like beasts of burden. In Russia between 1855 and 1861, there were almost 500 peasant uprisings, and not until 1905 did farm workers gain even theoretical equality before the law. As late as 1910 German agricultural laborers still had to work an 18-hour day and were treated like slaves.

In England farm laborers were free, but freedom held little meaning for them. To stay alive their wives and even six-year-old children often had to work. Girls of 10 and 11 sometimes labored in harvest gangs 14 hours a day for a pittance of seven pence. When British farm workers tried to form unions, they were jailed.

HUNGER AND EXHAUSTION:
THE FACTORY'S TOLL

The industrial system that Käthe Kollwitz knew produced for its workers little more than long hours and low pay on the job, poor food and sickness at home. Behind her sketches of an exhausted laborer at the end of the day *(above)* and factory workers eating bowls of soup *(right)* were bitter facts: one third of London's workers died on public charity; some 80,000 Viennese had no homes. But the voices of protest were being heard. Charles Dickens' novels scoring the injustices of capitalism were among the most widely read in the world. Charles Kingsley's *The Water Babies*, the story of a tiny chimney sweep, led to a law protecting children from being pressed into such brutal work.

DESPAIR, SQUALOR
AND ESCAPE

For the factory worker life was an endless cycle of squalor and despair; his only escape was in drink. In 1850 industrial Manchester had 400,000 people and 1,600 saloons; on Saturday night, it was said, "drunken women by the hundred lay about . . . in the mud." Käthe Kollwitz drew men roistering in a dingy Hamburg tavern *(left)* and she noted in her diary: "Frau Pankopf was here. She had a black eye. Her husband had flown in a rage. . . . For the woman the misery is always the same. She keeps the children . . . the man drinks."

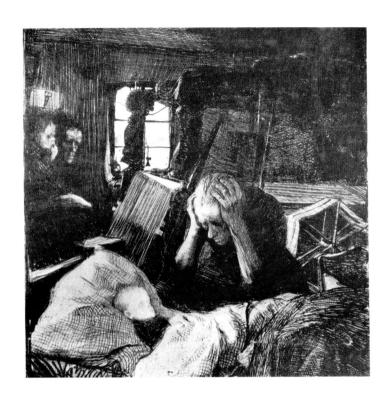

SICKNESS, DESPAIR
AND DEATH

Disease and early death were the constant companions of the masses crammed into Europe's cities. As the population multiplied around the factories, the overcrowded towns choked on their own dirt. "Filthy river, filthy river," one poet saluted the Thames; in the summer of 1858 its stench grew so vile that there was talk of moving Parliament elsewhere. Infectious diseases flourished; in 1850 they caused 94 per cent of all deaths in Europe. In Manchester, one of Britain's leading industrial cities, thousands lived in dank cellars that were a breeding ground for disease; most children born in these hovels never reached adulthood. Käthe Kollwitz had ample occasion to picture the misery of a parent watching over a sick child *(above)* or a husband mourning a dead wife *(right)*. "All my life," she wrote, "I carried on a conversation with death."

MOUNTING TENSION

By the late 1800s labor reforms were appearing: unions had been legalized almost everywhere, there were 30,000 credit unions, and labor was becoming militant. Workers met secretly to plot their moves *(far left)*, marched in demonstrations *(above)*, and even stoned the elegant mansions of their landlords and factory owners *(left)*. By the time Karl Marx died in 1883, his disciples were taking over budding trade union movements and converting them into weapons for class war. In the same decade labor won significant victories in the bloody Belgian miners' strike and the London Dock Strike.

A CALL
TO REVOLUTION

"Man must live for something better!" cried the Russian revolutionary Maxim Gorki in 1902. A character in Emile Zola's novel *Germinal* saw "the red vision of revolution that . . . at the end of the century would carry everything away."

Yet even as workers acquired the power and courage to revolt, they lost much of their need and desire to do so. Trade unions, the ballot and a greater share in rising productivity began to ease their lot. Nation after nation passed laws improving wages, hours and working conditions. Only Russia pursued old, dangerous patterns of repression. Marx, who envisioned the rich getting richer and the poor poorer, was only half right. The poor were also gaining: real wages in industrial Europe rose 50 per cent between 1870 and 1900. In this sense it had been an era of progress after all—but at a sad cost in human suffering.

5

NATIONS MADE BY "BLOOD AND IRON"

Not in memory had there been a reception quite like it. On that April day in 1864 half a million wildly cheering people lined the streets of London to greet a hero from abroad. Wearing his famous red shirt and his South American poncho, he rode in a carriage at the head of a parade of workers and trade union officials. His fair hair and beard were long in the manner of the Prophets. There was no more romantic or popular figure in all of Europe. His name alone was magic: Giuseppe Garibaldi.

The people of London were welcoming this almost legendary Italian because he had fought for freedom. Four years before he had led a gallant band of volunteers in support of a popular uprising in Sicily, then one of the many separate Italian states ruled by despots or by foreigners. The resulting emancipation of the island from the Bourbon dynasty was the first step in achieving Garibaldi's dream of a single Italian nation.

His campaign to unify the people of Italy—who had been divided among a number of unstable, hostile principalities since the fall of Rome more than a thousand years before—was launched in a spirit of untainted idealism. And although the ultimate attainment of his objectives resulted as much from the work of politicians and statesmen as from his efforts, it was Garibaldi's bold adventure in the name of freedom that captured the imagination of the European people. He had become a folk hero of epic proportions, a symbol of the desire for nationhood that was emerging throughout the subject lands of Europe. This striving for national freedom and integrity, despite all of nationalism's later distortions, remains one of the great romantic movements of history and one of the dominant themes of the Age of Progress.

"Man is born free and is everywhere in chains," Jean Jacques Rousseau had written in *The Social Contract*, published in 1762. He had in mind both national and personal freedom, for like other liberal observers of the 18th and 19th Centuries, he considered the rule of foreign monarchs over people of other tongues and racial antecedents a moral affront as great as any crime against the rights of man. Yet, even after the Congress of Vienna in 1815, which redefined Europe's national boundaries

THE IRON CHANCELLOR, *Prussia's Otto von Bismarck (left) used nationalist sentiment to unify Germany. The resulting empire, which he dominated for 20 years, became the strongest political and military force in Europe.*

in the wake of Napoleon's conquests and defeats, most of the continent's people were still bound in dynastic chains.

Dreams of nationhood nonetheless persisted. In the 1820s intellectuals and students flocked to join new nationalist movements along with other less articulate patriots. Many joined secret societies dedicated to fomenting revolt: the Carbonari in Italy (whose symbol was black charcoal that glowed until it burst into a bright flame), the Masonic lodges in Spain, the National Patriotic Society in Poland and groups of self-styled liberals in Germany who often met in so-called reading clubs. A manifesto issued later by nationalist leaders in London epitomizes the idealistic spirit of the times: "Everywhere," they declared, "Royalty denies national life . . . . Revolution alone can resolve the vital question of the nationalities, which superficial intelligences continue to misunderstand, but which we know to be the organization of Europe. It alone can give the baptism of humanity to those races who claim to be associated in the common work and to whom the sign of their nationality is denied; it alone can regenerate Italy to a third life, and say to Hungary and Poland 'Exist!' It alone can unite Spain and Portugal into an Iberian Republic; create a young Scandinavia; give a material existence to Illyria; organize Greece; extend Switzerland to the dimensions of an Alpine Confederacy, and group in a free fraternity and make an oriental Switzerland of Servia, Roumania, Bulgaria and Bosnia."

Thus were the theories of nationalism nourished by visionaries. Later in the century these dreams were played out in reality. The nationalist movements came to take three recognizable forms: the overthrow of a foreign ruler, as in Greece, where Turkish rule was finally cast off in 1829; the replacing of a "legitimate" native monarch by a popular government, as was attempted in France in 1830 and 1848; and the unification of several in-

dependent states, as in Italy and Germany during the 1860s.

The first of the nationalist causes to attract the attention and sympathies of European idealists was the struggle of Greek patriots to free their country from the rule of the Ottoman Turks. This battle for independence had great emotional impact among Europeans, in part because the savage rebel peasants of the Morea district in southernmost Greece were represented to the popular European mind as the descendants of the ancient, freedom-loving Greeks. Added to that bit of lore was the tragic, and romantic, death of the famous English poet Lord Byron, while attempting to aid the rebels. Altogether, the various elements of the Greek war for independence did much to popularize the nationalist cause.

In the wake of Greek independence, a wave of rebellion swept Europe in 1830. Although no other country except Belgium gained freedom or representative national government in these uprisings, they were a sure sign that nationalism and popular sovereignty were ideals that had found permanent places in the European mind; eventually, given the right time and the right leaders, they would triumph.

During this period one of the most ardent apostles of nationalism was the Italian Giuseppe Mazzini. In 1831 he formed the Brotherhood of Young Italy and planted the idea of a republic in the minds of many men, among them Garibaldi. Mazzini saw the nation-state as "the totality of citizens speaking the same language, associated together with equal political and civil rights in the common aim of bringing the forces of society . . . progressively to greater perfection."

Mazzini's views were romantic and democratic, and, in the light of future developments, quite unrealistic. But the enthusiasm fanned by him and other idealistic revolutionaries—such as the Hungarian Louis Kossuth, who spoke out passionately for a Hungary free from Austrian rule—was inflammatory. Suddenly, in 1848, all Europe exploded with revolutions.

Just as in the revolutions of 1830, the impetus came from the French, who deposed the "Citizen King" Louis Philippe. University students in Austria and Germany and patriots in Hungary and Italy tried to follow the example of their French brothers—with the result that hundreds of young people were either massacred or imprisoned. The revolutions ended in some cases with concessions in the direction of more representative government. More importantly, the revolutions, though generally unsuccessful, showed that nationalism as an ideal could not be plowed under and forgotten.

But with the dawn of the Age of Progress at mid-century another kind of man—a realist instead of a romantic idealist—appeared on the scene to play a different kind of role in the process of uniting Europe's peoples into nation-states. There was Francis Deák in Hungary, Adolphe Thiers in France, Count Camillo de Cavour in Italy and Otto von Bismarck in Prussia. Of these, two stand out as dominant figures: Cavour and Bismarck. Cavour united Italy by reconciling an existing people's revolution with his ambition to extend the rule of his king over the whole Italian peninsula. Bismarck united the German-speaking peoples by a combination of force and attractive social legislation.

Italy in the 1850s was a conglomeration of states and dependencies. In the north the Austrian Empire controlled Venetia and Lombardy. Piedmont in the northwest and the island of Sardinia formed the independent Kingdom of Sardinia-Piedmont under Victor Emmanuel II of the Italian House of Savoy. The Papal States, ruled from Rome, spread across the middle of the peninsula. South of them, comprising half of all Italy, lay another independent kingdom, that of Naples and Sicily, or the Two

Sicilies, ruled by an Italian Bourbon, Ferdinand II. There were also the Grand Duchy of Tuscany, centered on Florence, and the small duchies of Parma and Modena, ruled by junior branches of the houses of Habsburg and Bourbon.

This jumble of loyalties and antagonisms looked like the very pattern of opportunity to Cavour, who had become Prime Minister of Sardinia-Piedmont in 1852. He was a rather colorless aristocrat, with small round spectacles and a beard that barely fringed a deceptively candid face. Though apparently without guile, he was charged with ambition: he was willing to take any risks and to play ruthlessly with emperors, kings and commoners to unite the whole Italian boot under his sovereign, Victor Emmanuel. To gain Piedmont a position of importance in European politics, he had sent its troops to fight with the British and French against the Russians in the Crimean War of 1854, and had secretly approached Napoleon III, urging him to join the Piedmontese in driving the Austrians from Italy—all moves designed to clear the way for his king to claim the throne of a united Italy.

Until 1859, despite Cavour's maneuvering in the north and despite sporadic uprisings in the Two Sicilies, there seemed little real hope for Italy's development as a nation. Then, as if responding to the call of fate, Napoleon began to listen to Cavour's pleas to aid the Mediterranean land where the Bonapartes had originated. In the spring of 1859 the French Emperor declared war against Austria and French troops began to stream through the Alpine passes. There followed two great battles, Magenta and Solferino, in both of which the Austrians were routed by the French and Piedmontese allies. Thus Cavour's master stroke of international politics succeeded: the Austrians were driven out of Lombardy (which was ceded to Piedmont) and never again were the Habsburgs able to put a potent force into the field south of the Alps.

THE UNIFICATION OF ITALY *began when Sardinia-Piedmont and France defeated Austria in 1859 and the former recovered Lombardy. The next year Tuscany, Parma, Modena, Romagna, the Kingdom of the Two Sicilies and most of the Papal States joined the federation. Venetia was acquired in 1866; Rome was annexed in 1870 when French troops supporting the Papacy were withdrawn after the start of war between France and Prussia.*

With the defeat of the Austrian army a new wave of revolts swept over Italy. In April 1860, the people of Sicily rebelled against their Bourbon masters once again. The eruption came in Palermo, and one of its leaders was a plumber, Francesco Riso. The homemade bombs that the rebels hurled at the Bourbon soldiers failed to explode, Riso himself was killed and his revolution was soon over. Although it failed pathetically, Riso's uprising underscored the revolutionary state of affairs that had long existed in Sicily. This inflammable situation was well known to Giuseppe Garibaldi, the professional soldier-democrat who had already supported popular uprisings in South America as well as in his native land.

On the night of May 5, 1860, Garibaldi's romantic band of volunteers, The Thousand, sailed from Genoa armed with little more than a fervent belief in freedom (but with Cavour's secret approval). When they landed in Sicily, they were greeted at first with suspicion. The Sicilians were less interested in a united Italy or nationalism than they

were in satisfying such simple wants as the abolition of the salt tax, a reduction in the price of bread and a chance to work. Soon, however, their desire to overthrow their corrupt government merged with the romantic nationalism of Garibaldi. Peasants came on foot and on horseback, armed with a variety of weapons: pitchforks and blunderbusses, pruning hooks and ancient cutlasses. In the streets of Palermo women, children and priests joined in building the barricades. In the battle that followed, Palermo was liberated from Bourbon oppression, and Garibaldi moved on to seize the city of Naples.

From the north of Italy, Cavour watched intently the Sicilian uprising and Garibaldi's successes. The invasion of Sicily by The Thousand suited his purposes admirably—but as a popular movement, it created certain dangers to stability. He shrewdly analyzed the situation: "If the insurrection is put down, we shall say nothing; if it is victorious, we shall intervene in the name of order and authority." Cavour was as good as his word; after Garibaldi's victory he disclaimed any association with the rebels and deplored the threat to established order.

After achieving successes at Palermo and Naples, Garibaldi had planned to march on Rome and proclaim the unification of Italy. Cavour, however, had no intention of letting Italy be united by popular forces; cleverly and cautiously he moved to thwart Garibaldi's march. By manufacturing various "incidents" along the border of the Papal States, he was able to justify sending his Piedmontese troops into those states on the pretext of saving them from "revolution."

At the Volturno River south of Rome, Cavour's troops stopped Garibaldi. With disappointment but unflagging courage, Garibaldi agreed to halt and recognize Cavour's monarch, Victor Emmanuel, as Italy's sovereign. Guided now by Cavour alone, the Italian states voted one by one to become members of the new Italian nation. Within a year, on March 17, 1861, most of the peninsula was united under the Piedmontese King. In 1870 the ultimate goal of complete unification was achieved with the addition of Rome, the ancient capital.

But it must be added that during the late 1860s, when Victor Emmanuel's city of Turin was the capital, the newly unified nation of Italy seemed a far cry from the republic dreamed of by liberals like Mazzini and Garibaldi. Cavour had used their love of freedom and the accident of a popular revolt in Sicily to establish a conservative, if not autocratic, state. Nine tenths of the peasant population had no idea of the meaning of the new nation. It seemed to them only that the Piedmontese were the new masters of Italy. And in fact they were, imposing their king and their constitution on the rest of the peninsula. Yet at least there *was* a constitution (which gave the vote to 600,000 landowners in a population of 27 million), and at least there *was* a king, who became increasingly the true symbol of Italy's long-sought unity.

After 1860, with the appearance of other new na-

tions such as Romania, which had recently won virtual independence from Turkish rule, the map of Europe began to take a different shape. The dreams of the romantic poets and revolutionaries were being fulfilled: ancient ties of blood and tradition were being affirmed. But a curious new factor was undermining the romantic aspect of nationalism. Cavour's successful exploitation of a popular uprising in Italy inspired men in other countries to ride the wave of national aspirations to a goal not of freedom, but of power. These men were a handful of conservative, practical politicians like Cavour who, determined to preserve the established order, saw the popular urge toward nation-states as a political force that should be taken advantage of rather than resisted. As a result of their balanced policies, which combined age-old ideas of sovereignty with the new concept of *Realpolitik*, the politics of reality, modern states began to emerge in Europe that possessed both crowns and parliaments, liberal institutions and ancient prerogatives.

The new man of the age was not the romantic hero but the statesman, a politician who stood somewhere between the established hierarchy and the people. He came to be recognized as the most important man in Europe—if only because he could claim to understand the "reality" of the fluid situation that then obtained within and among the European states. Because he willingly sacrificed ideals to achieve his ends, he developed a reputation for unscrupulousness. Admitting this, Cavour sighed, "If we did for ourselves what we do for Italy, we would be great rascals."

Perhaps no one in the Age of Progress was more adept at the new game of *Realpolitik* than the Prussian, Otto von Bismarck. In the years from 1862 through 1870 he created what has often been called a German Reich, but what was, in reality, a Prussianized Germany.

In 1860, Germany consisted of some 39 separate

states ruled by princes, dukes and petty kings in which liberal or representative institutions played little part. These states were joined in a loose confederation, with Prussia and western Austria as the dominant members. As it had elsewhere in Europe, the emotional pull of a national heritage had come to the surface in the German states during the revolutions of 1848.

Though that chapter of German history was regarded by liberals as perhaps the nation's finest hour, it held a lesson of quite another sort for Otto von Bismarck, architect of the German Reich. Bismarck was the very antithesis of such rebellious romantics as Garibaldi of Italy, and he viewed with scorn the ideal of popular sovereignty. He came from a family of aristocratic Prussian landowners. Tall, erect and broad-shouldered, he was a man of great physical energy and soaring ambition, with an enormous appetite for food and drink. Driven by a desire to dominate others, he had a conservative's contempt for "tedious humanitarian babblers," and a penetrating intelligence unhampered by any excess of scruple. He made his attitude clear in a famous statement that foreshadowed the policy he was soon to employ with brilliant success: "Not through speeches and majority decisions are the great questions of the day decided . . . but by blood and iron."

When he was appointed Minister-President of Prussia in 1862 by the Hohenzollern ruler, King Wilhelm I, Bismarck had three major goals: to force Austria, the most powerful member, from the German Confederation; to establish Prussia as its leading state; and to make Berlin rather than Paris the diplomatic center of Europe. The achievement of these goals was to be grounded on the fulfillment of the German dream of nationalism.

The key to Bismarck's success in realizing his ends was the Prussian army, and he deliberately provoked wars to implement his policy. In 1864 he instigated a conflict with Denmark over the occupation of the duchies of Schleswig and Holstein, a dispute in which Austria stood by as an indulgent friend. When Austria finally awakened to the possibility that her ambitious ally might become too powerful, the two states entered into the swift and bloody engagement known as the Seven Weeks War. The Prussian army, equipped with such new weapons as breech-loading rifles and cast-steel cannon, and for the first time using the railroads to deploy troops, defeated the Austrians handily at Königgrätz, near Prague.

With Austria humiliated and effectively excluded from the affairs of the German states, Prussia became the leading member of a new North German Confederation which was formed by Bismarck in 1867. It included all the states north of the Main River, leaving only Bavaria, Baden and Württemberg among the larger German principalities to pursue their own destinies.

The stage was now set for the final flowering of Bismarck's ambition—Prussian pre-eminence among the nations of Continental Europe—and this could most quickly be achieved by force. In the early 1860s the Prussian army had been modernized; its ranks had been swelled by a system of universal military conscription and its artillery brought up to the standards set by the American Civil War. Few Europeans realized that Prussia had become an armed camp—a country possessed by its army. With England and Russia resting unconcerned on either horizon, only Louis Napoleon's unstable Second Empire stood directly in the way of Prussia's plan for European dominance.

In 1870 Bismarck found an excuse to launch his new army against France. Because the Habsburgs had been driven out of Spain by revolutionaries and there was no acceptable heir to the Spanish throne, an international dispute developed. Spain's suggestion that a Hohenzollern prince (distantly

NORTH
SEA

N

THE
NETH
LANE

BELGIUM

Sedan **X**
Decisive battle
of the
Franco-Prussian
War 1870

FRANCE

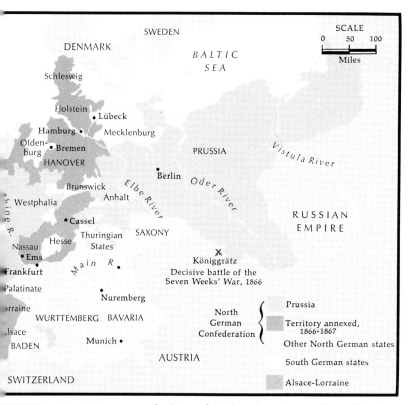

THE FORMATION OF GERMANY *under Bismarck is charted on the map above. After de-feating Austria in 1866, Prussia organized itself and other states north of the Main River into the North German Confederation. This group joined with the south Ger-man states to fight France from 1870 to 1871. In Louis XIV's palace at Versailles they proclaimed themselves one empire and acquired Alsace and Lorraine after their victory.*

related to Wilhelm I) be given the crown met with opposition from France, and Wilhelm and Louis Napoleon's ambassador met at the German spa of Ems to discuss the problem. After their conversa-tion the Prussian King sent a telegram to Bismarck in Berlin explaining that he had rejected a French request that he refuse to support the Hohenzol-lern candidate. In giving the telegram to the press, Bismarck changed it slightly to make it appear that the King and the French ambassador had exchanged insults. "This will have the effect of a red cloth upon the Gallic bull," Bismarck said, and he was correct. Coupled with a French desire to curb rising Prussian power, the telegram, known to his-tory as the Ems Dispatch, quickly produced the desired result. Within six days—on July 19, 1870—France declared war on Prussia.

At the outbreak of the Franco-Prussian War, the French army was filled with confidence. It had a new breech-loading rifle which outranged those of the Prussians, and Napoleon's generals assured him that their soldiers were ready "to the last gaiter button." They had not, however, made a careful study of the deadly steel cannon of the Prussians, or of the German railroad system which permitted the swift movement of troops, or of the efficiency of the Prussian staff.

From the start, the French campaign was a dis-aster, culminating in the decisive battle fought on September 2, 1870. On that day, near the small frontier fortress of Sedan, an overwhelmingly large German force surrounded a French army of nearly 90,000 men under Marshal MacMahon, a French-man of Irish descent who had borne the title "Duke of Magenta" since the Italian campaign of 1859. The French relied heavily on splendid but out-dated cavalry charges, and the artillery of the Prus-sians quickly cut them to pieces. Among the most senselessly heroic charges was one led by the Mar-quis de Gallifet. As the tattered remnants of his command straggled back to the French lines, Gallifet's superior ordered him to attack again. "As often as you like, *mon général*," replied the gallant marquis, "so long as there's one of us left."

Napoleon, who had taken command at Sedan, surrendered and on the following day was confined in the fortress of Wilhelmshöhe at Cassel in Germany. The French Second Empire had ended; a German one was beginning. But the city of Paris still fought on under a hastily assembled provisional government of national defense. Besieged by the Prussians, it held out until near-starvation forced it to capitulate at the end of January 1871.

On January 18, while the siege of Paris was still in force, the German princes under Bismarck's persuasion declared Wilhelm I of Prussia the first kaiser. For the French it was a supreme humiliation that the German nation-state, the Reich, was officially proclaimed in the Hall of Mirrors at Versailles, once the palace of Louis XIV and now Bismarck's headquarters. Ten days later the Germans entered Paris. With bands playing, they marched along the Avenue de la Grand-Armée and passed under the Arc de Triomphe, symbols of the might of France under Napoleon Bonaparte.

The German nation that emerged from the Franco-Prussian War was not the tightly unified, liberally oriented country that the romantic idealists had hoped for. Its acceptance by the German people lay in the twin fact that it appealed to nationalists because it offered some measure of unity, and to liberals because certain liberal reforms had been instituted, among them universal suffrage. If the liberals had not gained complete popular control of the government, they had at least gained some social progress—as well as something that could be called Germany.

Although the new German Empire was ostensibly a federation, in fact Prussia was dominant. In the words of the modern historians Ernest Knap-

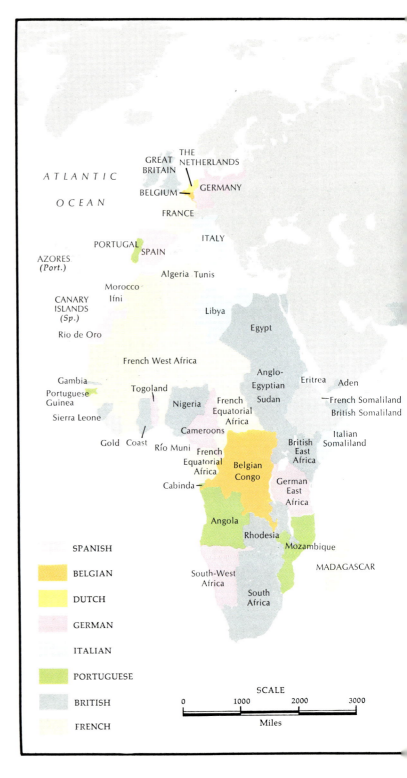

SPANISH

BELGIAN

DUTCH

GERMAN

ITALIAN

PORTUGUESE

BRITISH

FRENCH

SCALE

0   1000   2000   3000

Miles

THE STRUGGLE FOR EMPIRES *in Africa and Asia reached its peak during the Age of Progress. The newly formed nations of Germany, Belgium and Italy sought rich territories to help them match the older empires of Portugal, Spain, the Netherlands, France and Great Britain, seen on the map above. The fierce colonial rivalries among these nations helped start World War I.*

ton and Thomas Derry, "the Prussian monarchy, the Prussian bureaucracy, the Prussian army and above all the Prussian Minister-President, Bismarck, in his new role as imperial chancellor, exercised the decisive power." The upper house of the bicameral legislature consisted of members nominated by their individual regional governments, and was legally the seat of authority within the Empire. Thus it easily was able to undercut the power of the popularly elected lower house. Furthermore, Prussia wielded an effective veto in either house which could kill any measure. And, lest the popular will of the German people should nonetheless be expressed too freely, executive power was the sole prerogative of the emperor and of the imperial chancellor, who was entirely responsible to the emperor.

By such systems throughout Europe was the freedom of the individual—which had originally been implicit in nationalism—submerged in the triumph of the state. And this unforeseen trend away from popular sovereignty undoubtedly contributed to the collapse of the dream that international peace might result from the creation of European nation-states. In most European countries after 1871 the trend was no longer in the direction of democracy, but rather toward the extension of statist power. Feelings of racial unity, instead of simply being shared and relished, became something to defend. People were not merely Germans or Italians or French; they were belligerently so. Many countries took a lesson from Prussia's powerful army and began to use universal military conscription to build strong forces of their own.

To justify the growing violence of the era, Darwin's theory of natural selection was cited; twisted into a new context, Darwinism seemed to lend scientific support to the social doctrine of "survival of the fittest"—the idea that some nations were naturally stronger and better qualified to dominate than others. This sort of reasoning contributed to two developments that sparked aggressive nationalism and that were the very antithesis of the original concept of national spirit.

One of these developments was the rise of imperialism, which denied a people's right and capacity to govern themselves. Having consolidated their rule on their home continent, the nation-states of Europe were now ready to extend their domination abroad by acquiring colonies and protectorates in the backward countries of Asia, Africa and the Pacific.

European imperialism, to be sure, grew out of many factors. Economically, expanding industries and improved transportation stimulated the search for new markets, new sources of raw materials and new fields for investments. Interest in overseas expansion was also spurred as increasing numbers of missionaries to undeveloped countries argued that those countries would be better off under European protection and civilization than they were under their own barbaric chieftains. But the roots of imperialism were largely nationalistic—a drive for increased national power and prestige through the extension of empire. In 1871, Great Britain had greater colonial holdings than any other European nation; to France, Germany, Italy and other states the possession of colonies became a symbol of greatness, of having "arrived" on the international scene. By 1914 they had joined with the British to spread European control over more than half of the land surface of the earth. The Boer War of 1899 was the only extensive conflict to result from the grab for overseas possessions, but as national patriotism became more frenzied and more selfish, the quest for territorial domains led increasingly to the brink of war.

A second development that fostered aggressive nationalism was the idea of superiority of race, which led to so-called "pan" movements among various national groups toward the close of the 19th Century. The Pan-German movement, for example, propounded the superiority of all things German and sought to unite Germans everywhere linguistically, culturally and politically. Its followers advocated expansion overseas of the German "master race," and pledged themselves to combat all forces that stood in the way of German national development; they also envisioned a central Europe ruled by a Greater Germany, which was to be formed by forcibly incorporating the whole or parts of adjacent nations. "We want territory even if it be inhabited by foreign peoples, so that we may shape their future in accordance with our needs," wrote Ernst Hasse, who in 1894 founded the Pan-German League. Similarly, Pan-Slavs dreamed of uniting the Slavic-speaking peoples into a federation centered in Russia and exerting its influence throughout the world.

As Europe entered the 20th Century, the nationalism which had begun as a romantic movement in the direction of freedom and racial harmony had foundered on international aggression and racial antagonisms. Nevertheless, nationalism, with all its perversions, must ultimately be viewed as a progressive force: as a result of nationalism, most of the European continent had been transformed from dynastic systems into self-confident national units of dynamic power and sound economy. These nations in turn had reshaped the map of the earth as their drive for imperialism spread European domination—and the achievements of Europe's materialistic civilization—throughout the world.

The search for a new social order went on. Now, however, instead of seeking to alter governments by reason and constitutional means, people were turning to direct action and violence. Forces were soon to be released which had been held in check ever since men had decided that the mind should govern the emotions.

PLAYING SOLDIER, *two Italian boys gaze at portraits of Garibaldi (right), and King Victor Emmanuel.*

HERO FOR AN AGE

In the early 19th Century, a wave of nationalism swept Italy. Tired of domination by assorted princely despots and foreign powers, the Italian people wanted, in the words of one patriot, a "New Italy, a United Italy, the Italy of all Italians." For this they needed a leader, and in Giuseppe Garibaldi they found him. This flamboyant figure, with his impassioned speech and messianic appearance, embodied the essence of the nationalist ideal and he became a symbol of an age in which men fought for self-determination. Envisioning the Piedmontese King, Victor Emmanuel, as ruler of all Italy, he rallied an improbable band of volunteers and, in a series of colorful campaigns, helped to forge the Italian nation.

A REBEL ON THE RUN

It was while working as a seaman that Garibaldi first heard the whispers of popular revolt in his native Italy. At that moment, he later wrote, he felt "as Columbus must have when he first caught sight of land." He immediately joined the insurgent group, Young Italy, and within a few months was condemned to death for inciting revolution. He escaped to South America where, while waiting to return to his homeland, he fought for 12 years to establish republics in Brazil and Uruguay.

At last in 1848, Italian rebellion broke out again in Milan, and Garibaldi sailed back to help the insurrection. Although relentlessly hunted by French and Austrian armies intent on protecting their interests in Italy, he led a series of uprisings—and slowly but surely began to win popular support.

THE FACE OF THE HERO, *painted while Garibaldi was in Uruguay in 1845, shows the piercing eyes and prophetlike beard that so stirred the Italian spirit. The sombrero he wore was to become an integral part of his uniform when he returned to Italy.*

FLEEING CAPTURE, *Garibaldi, disguised as a peasant, carries his wife Anita through the swamps of Comacchio, north of Rome, in 1849. At his heels were Austrian soldiers sent to suppress the revolution. Garibaldi's guide was one of his own officers.*

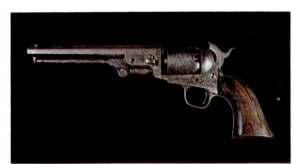

GARIBALDI'S PISTOL *was a Colt .36 Navy model made in 1851.*

A CLANDESTINE MEETING *of Garibaldi's followers is held by candlelight in an Italian village. Groups of patriots like these gathered secretly throughout the country to plot uprisings against their governments. They risked exile or imprisonment if captured.*

THE MAN IN THE RED SHIRT

Although Garibaldi's initial attempts to liberate Italy were unsuccessful, his campaigns were extremely colorful and his name and the red shirt he always wore soon became symbols of the struggle for Italian unity.

In 1860 he devised a master plan. The liberal King Victor Emmanuel of Piedmont commanded a powerful army in northern Italy; if Garibaldi could win a decisive victory in the south, the two forces could then merge and crush their enemies between them. Immediately Garibaldi made ready to attack Sicily. He gathered a thousand volunteers armed with obsolete rifles and rusty bayonets; with this small force, most of whom had never seen combat, he invaded the island and boldly confronted the 24,000 Bourbon troops of King Francis II of Naples.

THE TRICOLOR FLAG of a united Italy is carried by Garibaldi into Marsala on the west coast of Sicily. He wears his famous red shirt, a uniform he adopted in South America, where the shirts were worn by butchers to hide the marks of cattle blood.

THE EMBARKATION of Garibaldi's troops from Genoa marked the beginning of his strongest bid for Italian unity. The volunteers sailed to Sicily in two crowded steamships; during the rugged 400-mile journey most of the men became seasick.

A DRESS SWORD used by Garibaldi bears the inscription "Viva l'Italia unita"—"Long live a united Italy."

A CRUCIAL VICTORY

Garibaldi's first battle in Sicily matched his men—"The Thousand," they were called—against more than 3,000 Bourbon troops firmly entrenched near the village of Calatafimi. Again and again he attacked, and each time was thrown back. Then, he was suddenly hit with a large stone; exultantly he cried out that the Bourbons were out of ammunition. The hand-to-hand fighting that followed lasted for hours—but when it was over the rebels held the field. The victory inspired the people of Sicily and they joined the fight by the thousands. Within six weeks the entire island was liberated.

LEADING THE CHARGE, *Garibaldi (top right) urges on his outnumbered red-shirted volunteers against the leather-hatted Bourbon troops during the crucial*

battle of Calatafimi in 1860.

THE WARRIOR ON HORSEBACK, *Garibaldi still wore his South American sombrero and a colorful scarf. A young soldier in his army once described Garibaldi astride his horse as "one of the noblest things in art or nature."*

GARIBALDI'S SADDLE, *an elaborate model he used only on ceremonial occasions, bears an emblem of Victor Emmanuel's House of Savoy on its twin gun holsters (left) and the King's personal insigne on its blanket.*

A KING IN HIDING, Francis II, monarch of the Two Sicilies, sulks in a cellar while Garibaldi marches into Naples. Francis, one of the last monarchs in divided Italy, personified the despotism Garibaldi fought to extinguish. His queen is seen praying at right.

JOINING THE NEW NATION, Tuscan officials arrive in a delegation to offer their allegiance to Victor Emmanuel. Infected with the spirit of nationalism, the Tuscans had overthrown their ruler, Grand Duke Leopold, to side with Garibaldi's band of freedom fighters.

THE FRUITS OF TRIUMPH

Garibaldi's victory in Sicily gave the nationalist forces control of one-half of the population of Italy. King Victor Emmanuel had annexed most of northern Italy, including Tuscany and Romagna, leaving only Venice, the Papal States around Rome, and Naples still to be taken. According to plan, Garibaldi then crossed the Strait of Messina and marched up the boot toward Naples.

By this time his popularity was immense. He had proclaimed himself dictator of Sicily and Naples, but he still intended to hand over his conquests to Victor Emmanuel of Piedmont once the victory was complete. To prove this allegiance he occasionally donned the uniform of the Royal Piedmontese Army. Confident of victory, the Italian people rallied behind him enthusiastically, and soon his small band of volunteers multiplied into a large army. The demoralized Bourbon troops were routed in every skirmish. On September 7, 1860, Garibaldi triumphantly entered Naples without firing a shot.

GUERRILLA TURNED GENERAL, *Garibaldi stiffly poses in a Royal Piedmontese Army uniform. After he accepted his commission, he occasionally had to abandon his beloved red shirt for this less comfortable garb.*

A NATION DELIVERED

On the morning of October 26, 1860, Garibaldi stood in front of a tavern near the town of Teano watching the resplendent regiments of Victor Emmanuel's army file past. Garibaldi had been viewing the parade for hours when he heard the royal march and men shouting "The King is coming!" He mounted his horse and rode toward Victor Emmanuel waving his hat and shouting *"Saluto il primo Re d'Italia!"* —"I hail the first King of Italy!" The two men clasped hands, and at that moment Italy became a nation.

Garibaldi's example, and the new power of nationalism, quickly spread through Europe. It had been shown that men willing to fight and die could overthrow their rulers. Within 10 years the powerful Austrian Empire had declined and the strongest of the nation states, Germany, had emerged. Small wars between monarchies were soon replaced by larger wars between nations. Giuseppe Garibaldi, his personal struggle over, became an idolized and almost legendary figure in his own time.

A RICHLY EMBROIDERED HAT, *called a papalina, replaced Garibaldi's sombrero in his later years.*

6

THE FLAG
THAT FAILED

For two months in the spring of 1871 a people's government controlled Paris. It was called the Paris Commune, but it had nothing to do with communism and its members had by no means plotted violent revolution. They were not even exclusively working class. Although there were Marxists and socialists among them, there were also republicans and latter-day Jacobins, heirs to the democratic ideals of the French Revolution, which had given France its First Republic, 78 years before. The mixed peoples who made the Commune were united for one purpose: to create, once and for all, a government truly representative of the French people. "The governing classes," said Citizen Jean-Baptiste Millière, "are putrefying, and French civilization is forever lost if it any longer remains in the hands of this rotting oligarchy."

Behind the Commune lay years of discontent with an autocratic French government, but the immediate cause was the government's conduct of the Franco-Prussian War. In September 1870, less than two months after he had declared war on Prussia, Napoleon III crossed through the German lines at Sedan on the Meuse River to surrender himself and his beaten army. "Nothing remains for me but to give my sword into your hands," he wrote the Prussian king. But the people of Paris did not agree. Besieged by German troops, cut off from all supplies, reduced to eating horsemeat, zoo animals and even the rats in the sewers, they fought on. In February they received a second blow. The French National Assembly, newly elected to deal with Bismarck's surrender terms, agreed to let German troops occupy the city. Paris greeted the news with shock and dismay, met the Germans with closed shops and black-draped streets —and overturned the Assembly's authority.

The goals of the Commune were moderate: it simply wanted what working people had wanted for decades—better wages and working conditions, and enough government control of prices to assure them of some economic security. Nevertheless the reign of the Commune was marked by scenes of appalling violence. To the National Assembly, governing France legally from Versailles, the rebellious Communards were dangerous radicals to be put down at any cost. Government troops lined

JUBILANT REVOLUTIONARIES in Odessa, Russia, carry a portrait of the Czar in 1905, after forcing him to grant them a constitution. The next year the Czar reneged and smashed the popular movement with mass murders.

up Communard captives and rode along the line striking at them with the flat of their sabers. Communards retaliated by burning public buildings and assassinating the Archbishop of Paris. "Paris the beautiful," wrote an English reporter, "is Paris the ghastly, Paris the battered, Paris the burning, Paris the blood-spattered, now."

Between May 26 and June 2 the bloodshed reached incredible proportions. The conservative government of the Assembly, regaining control of the city, massacred Communards by the thousands, in some cases disposing of the bodies so hurriedly that some were buried alive. People living near one of the mass graves reported hearing groans in the night and seeing "in the morning a clenched hand . . . protruding from the soil." When it was all over, 20,000 or more Communards had been slain. The only reform the Commune had gained, one of its leaders ironically observed, was an end to night work for Paris bakers.

As a symbol, however, the Commune lived on—at least in certain places. Karl Marx, whose followers neither led nor inspired the rebellion, nevertheless saw its relevance to his cause, and used it. "Workingmen's Paris, with its Commune," he wrote, "will be forever celebrated as the glorious harbinger of a new society." When the Bolshevik hero, Lenin, died in 1924 his body was wrapped in the red flag of the Commune. Lenin, in fact, had spent a great deal of time studying the Commune. He admired it for its heroism and criticized it for its errors—one of the greatest of which, he pointed out, was the "unnecessary magnanimity of the proletariat."

Even if the leaders of the Commune were not ruthless enough, by Lenin's standards, they did unquestionably resort to measures of extreme violence. Steadily, and with increasing impact, the working-class masses were rising to the surface—but rising, unhappily, by means that were often violent. Revolt and rebellion gradually came to be not last resorts, but first ones, and for many men they were the only effective weapons of political and social protest. The Commune of 1871 uncovered hidden currents of anger and aggression around which the history of the 19th and 20th Centuries might well be written. Revolution, said Karl Marx—in a phrase peculiarly appropriate to an industrial civilization—was "the locomotive of all history."

Certainly that civilization was supplying plenty of combustible materials. Although the Age of Progress was an age of peace at the national level, it was also an age of unrest. There were no large-scale wars, but there were continual minor eruptions over a variety of issues. Chief among these issues were the glaring differences between the life of the rich and that of the poor, differences that the poor tended less and less to accept with equanimity. "What is it to me that there should be monuments, operas, café-concerts?" said one French worker who had neither the time nor money to enjoy them. And in Italy, Enrico Malatesta found plenty of willing listeners to his anarchist views: "Do you not know," he cried, "that every bit of bread they eat is taken from your children, every fine present they give to their wives means poverty, hunger, cold, even prostitution for yours?"

Even in countries like Britain, Belgium and Sweden, workers remained dissatisfied. In spite of rapid advances toward universal suffrage, universal elementary education and social reforms, the main problem was still unsolved: social inequality. The Industrial Revolution had, it seemed, unleashed a force it could not control. Despite its almost unlimited capacity for producing material wealth, it was apparently incapable of distributing that wealth equally through all the ranks of society. And that failure endangered the very fabric of society. It created the possibility of a better life for all men, while at the same time denying that good life to the working class who had helped to achieve it.

The nature of this danger was seen early, and in another context, by a most improbable man. In 1856, Czar Alexander II called together a group of Russian noblemen to deny the rumor that he was about to abolish serfdom. "But you yourselves are certainly aware," he warned them, "that the existing order of serfdom cannot remain unchanged. It is better to abolish serfdom from above than to wait for the time when it will begin to abolish itself from below." It was just such a rebellion from below that had produced the Paris Commune and that, as Europe turned into the 20th Century, was increasingly producing incidents of violence in many other countries. But the European Establishment in general, and Russia's czars in particular, refused to heed the warning.

The workingman's initial weapon was the strike, and the early years of the new century saw a wave of work stoppages. In 1902 Belgian trade unionists called a general strike to demand universal suffrage, and a year later Dutch trade unionists struck in protest over antistrike legislation. In 1904 a strike of Milanese workers spread over the whole country and for four days brought the nation to a standstill. In 1905 the Austrian socialist Viktor Adler, founder and leader of the country's Marxist-oriented Social Democratic party, called a massive strike to climax a campaign for universal suffrage; thousands of workers marched with clenched fists and red flags into Vienna's Mariahilferstrasse—and the Austrian government granted them their wish. In 1906 the French *Confédération Général du Travail* struck unsuccessfully for an eight-hour day and in the next few years there were other strikes among French miners of the Nord, dockers at Nantes and vineyard workers of the Midi.

But the protests of dissatisfied workers were mild affairs compared to the disturbances created by political extremists. The most widespread of these were the anarchists, and the most outspoken anarchist was Mikhail Bakunin, a giant of a man who spent most of his life rebelling against all forms of social order. Bakunin, a Russian aristocrat, resigned from the Imperial Guard in 1835 in protest over the Czar's autocratic methods. Thereafter he traveled through Europe allying himself with first one, then another revolutionary cause. He participated in the 1848 revolution in Paris, helping to depose the French monarchy, and in the same year turned up in Prague to encourage demonstrations of Czech nationalists. In 1849 he took part in a rebellion in Saxony, for which he was first thrown into prison and then shipped back to Russia—where he was imprisoned for six years more and then exiled to Siberia.

Bakunin escaped from Siberia in 1861, and arrived in London in the same year, toothless and aged almost beyond recognition, but still fiery of spirit. He recovered among friends and picked up where he had left off.

For a time Bakunin allied himself with Marxist socialism. He joined the First International Workingmen's Association in London in 1868, four years after its founding, but quarreled violently with Marx and a few years later was forced out. Unlike Marx, who wanted to seize control of the state in the name of the proletariat, Bakunin wanted to destroy the state entirely. He believed, like the Marxists, that the accumulation of private property was evil: "Property is theft," wrote Pierre-Joseph Proudhon, the radical thinker whom Bakunin acknowledged as his master. But Bakunin also believed, with earlier anarchists, that the only real ordering force in society was the individual man himself: governments were illusions.

To Bakunin this goal of extreme liberty justified extremes of violence. Ironically, his advocacy of violence was one of the reasons for his expulsion from the International—although violence later became a tool of Russian revolutionaries. He believed that destructiveness was creative, and therefore good, and he spoke of unchaining the lower echelons of society, the "solid, barbarian elements," to roll "like a raging avalanche devouring and destroying" their enemies. By the 1890s Bakunin's doctrine had gained thousands of adherents. Paris was full of anarchist periodicals with such provocative names as *New Dawn, Black Flag, Enemy of the People, People's Cry, The Torch, The Whip.* On May Day in 1891 anarchists in Clichy, a working-class suburb of Paris, put on a demonstration that was broken up by mounted police. Bomb-throwing became a favorite form of direct action. A street song of the day had this haunting refrain:

> *It will come, it will come,*
> *Every bourgeois will have his bomb.*

The anarchists' most dramatic acts of social protest were assassinations. Between 1881 and 1901, anarchist conspirators disposed of no fewer than five heads of state. In 1881, after seven tries in two years, the People's Will, a secret Russian terrorist society, threw two bombs against the carriage of Alexander II, blowing off his leg; the Czar soon bled to death. In 1894, Sadi Carnot, the President of France, was assassinated at Lyons by an Italian anarchist who professed to be avenging a French anarchist condemned to death for throwing a bomb in the French Chamber of Deputies. In 1898 the Empress Elizabeth of Austria was stabbed to death by an anarchist who had vowed to kill the first royal person he could find; he found Elizabeth walking to a boat on Lake Geneva. Two years later King Humbert I of Italy met his end at the hands of an anarchist assassin at Monza, and one year after that the American President,

William McKinley, was shot to death by an anarchist at Buffalo, New York.

Russia seemed to provide especially fertile soil for this sort of bloody activity. For more than half a century some of the country's best minds had preached the violent overthrow of Russia's autocratic government; in 1900 this agitation crystallized into a political movement, the Social Revolutionary party. The Social Revolutionaries were not Marxists, and they were not primarily concerned with the urban proletariat; their basic goal was a better life for the peasants and the socialization of the land. But their means to this end were violent. In the two years following the party's formation, Social Revolutionary terrorists killed the Minister of Education, the head of the Secret Police and a provincial governor. In 1904 the party plotted the murder of Wenzel von Plehve, Minister of the Interior and the hated instigator of many Jewish pogroms. And in 1905, in one of the most spectacular Russian assassinations, a bomb hurled by a Social Revolutionary reduced the Grand Duke Sergius Alexandrovitch Romanov, along with his carriage and horses, to "a formless mass of fragments about eight or 10 inches high."

At his trial the Grand Duke's assassin, a young man named Kaliaev, put into words the tensions that lay behind all the strikes and murders and violent protests that were occurring with increasing frequency all over Europe. "We are two warring camps," he told his judges, ". . . two worlds in furious collision. You, the representatives of capital and oppression; I, one of the avengers of the people." That such a statement could be made at the height of an age of material progress points to the great failure of those years of hope. Man's technical ability had far outstripped his moral advancement. A time had come when the people demanded to be listened to. The Russian government chose not to heed the demand, and the result gave

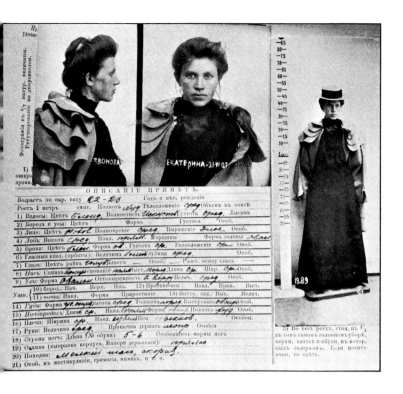

Europe a preview of the future. But Europe, ruled by a complacent bourgeoisie, paid little attention to the events in Russia in 1905.

The Revolution of 1905, like the Paris Commune of 1871, was a spontaneous popular outburst. There were no plans for it, and no inflammatory speeches by socialists or anarchists lit the fuse. The people had simply suffered enough, and protested. Lenin was as much surprised by it as Marx had been by the Paris Commune 34 years before, but like Marx he made use of it for his own ends. The Revolution began on January 22 in St. Petersburg. Less than three weeks before Japan had seized Port Arthur, marking one more defeat for Russia in the disastrous Russo-Japanese War—a war that Czar Nicholas II had hoped to use to rally public support for his government. Only a few days before, 13,200 workers in the Putilov engineering works in St. Petersburg had gone out on strike, but had been unable to gain the ear of their employers.

The strike had been organized by Father Gapon, a priest sent among the workers by the police to channel their protests into peaceful demonstrations. At Father Gapon's suggestion, the workers now decided to appeal directly to Nicholas, and prepared a petition: "We, the workers of the town

of St. Petersburg, with our wives, our children and our aged and feeble parents, have come to you, Sire, in search of justice and protection. We have fallen into poverty, we are oppressed, we are loaded with a crushing burden of toil, we are insulted, we are not recognized as men, we are treated as slaves who should bear their sad and bitter lot in patience and in silence." They asked Nicholas for a voice in the government, for a constituent assembly and "universal, secret and equal suffrage." They requested an eight-hour working day, the abolition of overtime, heated workshops and medical attention.

Shortly before noon on Sunday the 22nd, dressed in their best Sunday clothes, the people poured into the great square in front of the Winter Palace—more than 200,000 of them—to present their petition to the Czar. The event was in the nature of a holiday. Peasants had come to town in their brightest embroideries and parents brought their children along. Many of them carried pictures of the Czar, their "Little Father." At their head walked Father Gapon and two other priests, carrying icons. "God save our people," they sang, "God give our Orthodox Czar his victory." But the Czar did not hear them; he was not there. And the officials grew alarmed.

As the crowd pressed forward, the Imperial Guard, drawn up in front of the palace, opened fire, and the front ranks of the workers went down. After that, all was chaos. Cossacks rode into the mob, laying about them with knouts and sabers and finally using guns. "I have seen blood flow in streams on the hardened snow," telegraphed a reporter from *Le Journal* of Paris. "I have seen police agents, sword in hand, slash blindly about them. I have seen their revolvers used wildly against the crowd. I have seen whole companies of infantry discharging murderous volleys on the shrieking crowd. And on all sides the dead with the wounded falling upon them, and the horrible pell-mell, in which women and children covered with blood fall in the snow."

By evening perhaps 1,000 people had been killed and thousands more were wounded. Father Gapon, who had miraculously escaped with his life, sat down and wrote an open letter to the Russian workingman. "There is no Czar," he wrote. "Between him and the Russian Nation torrents of blood have flowed today. It is high time for the Russian worker to begin without him to carry on the struggle for national freedom." Thus ended the day that went down in history as Bloody Sunday, and thus began the Russian Revolution of 1905. It was a revolution against many things—against a huge and bureaucratic state, against brutal and oppressive poverty, and against the evils of uncontrolled capitalism.

The Industrial Revolution came to Russia late. It did not arrive in full force until the 1890s, when trade expanded tremendously and foreign investment capital flowed into the country in enormous quantities. A great coal-mining industry grew up in the Donets Basin and by 1891 the first tracks of the 5,800-mile Trans-Siberian Railroad were being laid. Moscow became a textile center and St. Petersburg a center for the manufacture of machines and machine tools. The Urals were producing iron, and Baku, petroleum; the Russian area of partitioned Poland was an important source of coal, iron and steel, and chemicals. By the end of the century Russia stood fourth in the world production of four basic commodities—iron, coal, oil and cotton—and its growth rate exceeded that of any other European country, including England and Germany.

Most of this industry was highly concentrated. By 1902 nearly half of Russia's industrial workers were employed in factories having labor forces of 1,000 people or more. As a result Russia faced, at

a late date, many of the social problems that England had faced in early Victorian times, France in the Second Empire, and Germany when Bismarck first came to power. Living conditions in urban centers were appalling. But unlike the governments of Western Europe, which attempted to correct such conditions, the Russian government did nothing.

As for the peasants, they were no better off than the industrial workers. Although the abolition of serfdom had freed farm workers from servitude, their condition was in some ways worse than before. The plots of land assigned to them for their own use were often the least fertile; moreover, they were not outright gifts, but had to be paid for—usually in installments over long periods of time. Also, while the peasants were technically free, in actuality they could not leave their villages without a passport and the permission of the village elders. Agricultural stagnation was common and starvation was chronic. In the 1890s a series of famines swept through rural Russia, adding to the peasants' misery.

Thus the Russian Revolution of 1905, coming exactly at the moment when industrial civilization was bestowing its first blessings on Russia, revealed widespread deficiencies in that civilization.

After the horror of Bloody Sunday in St. Petersburg, violence spread to other urban centers of Russia. By the end of the month nearly half a million workers were out on strike, and the streets of cities like Moscow, Odessa, Warsaw and Lodz were the scene of wild rioting and terrible retaliatory massacres. Political revolutionaries appeared from underground or returned from exile to organize the workers and direct their protests. Middle-class intellectuals and professionals, for once in sympathy with the workers, added their voice to complaints about the government. In March Nicholas reluctantly made a few halfhearted concessions which not very many people took seriously—including Nicholas himself.

In May the government suffered another blow which added to its unpopularity. Almost an entire Russian fleet—22 ships and 6,000 men—was destroyed by the Japanese Navy in Tsushima Strait, between Japan and Korea, plunging Russia's fortunes in the Russo-Japanese War to a new low. As summer approached the civil disorders grew more intense. Peasants joined the revolutionary cause, burning and looting the manor houses of rich landowners. Unrest spread to the armed forces where, in June, it burst into the open: sailors aboard the battleship *Potemkin*, pride of the Black Sea fleet, mutinied and killed seven of their officers, proclaiming "Down with the autocracy! Long live the Constituent Assembly!" In Geneva, Lenin, the exiled leader of the militant Bolshevik wing of Russia's Marxist party, the Social Democrats, took time out from his theoretical writings; he urged St. Petersburg workers to make bombs and arm themselves with guns and knives.

In August, Nicholas agreed in principle to the idea of a representative assembly, but the Duma, or parliament, that he proposed fell far short of the people's demands. It was a consultative body only, and its membership did not include the urban workers. Consequently the strikes went on. In October a colossal work stoppage paralyzed the whole city of St. Petersburg and much of the rest of Russia besides. Starting as a minor strike of printers who were demanding payment for setting punctuation marks, it spread to other trades and professions and touched off companion strikes in Moscow. Trains stopped running, banks closed, telegrams and mail went undelivered, lawyers refused to practice, bakers refused to bake, and the *corps de ballet* refused to dance.

The October strike had two important results. It was run by a council, or soviet, of workers—the first of many such soviets the nation was to see. Numbering at its height some 400 delegates, the St. Petersburg Soviet represented all shades of democratic opinion, from mildly liberal to radically revolutionary, and it spoke for a politically powerful unit of some 200,000 people; one of its first presidents was a young revolutionary named Leon Trotsky. The other important outcome of the strike was the October Manifesto, another promise from Nicholas. Although it subsequently proved to carry no more weight than his previous promises, it did go further; it established a democratically elected Duma with legislative powers, and it extended the basic civil liberties—freedom of speech, assembly and conscience—to everyone. Satisfied that Nicholas had finally capitulated, the moderates deserted the revolutionary cause—and the Revolution of 1905 had passed its climax.

But it was not, of course, gone from men's minds and desires. "The Revolution is dead, long live the Revolution!" exclaimed Trotsky. And in the penal colonies of Siberia, and the coffeehouses of Vienna and Zurich, intellectuals and professional revolutionaries continued to study and talk of the events of 1905. Lenin saw it as a great rehearsal. "The revolutions in Turkey in 1908, Persia in 1909 and China in 1911," he wrote long afterward, "prove that the mighty uprising of 1905 left deep traces, and that its influence, expressed in the forward movement of hundreds and hundreds of millions of people, is ineradicable."

There were lessons to be learned from this violence, smoldering on the edges of Western Europe. The rising power of the masses threatened the harmony of the existing social system. Even in countries like England, France, Italy and Germany the basic causes of the Russian Revolution of 1905 were present: the growing power of the state, discontented workers, class antagonisms. Although few heeded the lessons, and most people continued to believe in progress, the turn of the century was nevertheless a time of tension. There were sharp skirmishes between those who wanted change and those who wanted the world to stay as it was.

The pressure for change could be read in statistics and in events. In Germany, the number of votes for the Social Democratic party grew from half a million in 1877 to almost a million and a half in 1890; 22 years later the total was over four million. In Britain by 1914, class and party differences were so pronounced that the country stood on the brink of a general strike. It also faced the prospect of civil war: Ireland wanted home rule. The Irish had never recovered from the potato famine of 1846-1851, when nearly one million people died of starvation or disease, and nearly two million more fled the country. Those who remained behind fought a constant battle against absentee English landlords and poverty. Every day more tenants, unable to pay their rent, were evicted by the police, while angry mobs gathered to hurl stones, vitriol and boiling water at Her Majesty's servants. Ireland, in fact, was for England an almost insoluble problem.

But it was in France that the antagonisms below the surface of turn-of-the-century Europe erupted most violently. France was rocked in the 1890s by the Dreyfus Affair, an event which divided the country into two warring camps and plunged the nation into what a later premier, Léon Blum, called a "veritable civil war." At the heart of the Affair was Captain Alfred Dreyfus, the only Jewish officer on the French General Staff. In 1894 Dreyfus was found guilty of selling military secrets to Germany and was sentenced by a military court to life imprisonment on Devil's Island. Two years later evidence emerged that the culprit was another man, an adventurer named Major Ferdinand Walsin-Esterhazy. But the army refused to reopen the case, and even produced a number of forged documents that purported to demonstrate Dreyfus' guilt.

The Affair turned France into a battleground. Men argued in cafés, in the halls of state, in public print and in the bosoms of their families. The issue was less Dreyfus' guilt or innocence (eventually he was fully exonerated) than the question of whether an injustice to one man warranted an attack upon the institution of the army, especially when that army was a critical factor in France's relations with Germany. Around this issue clustered a series of satellite issues—of Church vs. State, Right vs. Left, Christian vs. Jew, Aristocracy vs. Common People, Monarchy vs. Republic.

The defenders of Dreyfus were only partly concerned with justice; they also wanted to discredit the forces of reaction—the royalists, clerics and upper echelons of the army who wanted to re-establish the ancien régime. Those who attacked him were defending that regime, and many of them were also anti-Semites. They resented Jewish revolutionaries like Marx and Trotsky, and despised Jews in general. When the anti-Semitic journal, La Libre Parole, started a fund for the widow of the man who had written the forged Dreyfus documents, subscribers accompanied their contributions with letters proposing that Jews be used as test targets for new guns and explosives, that they be blinded, poisoned, thrown into sewers.

The Dreyfus Affair left France a different country. It ended the power of the ancien régime and brought to power the petit bourgeois: except for three years, a Radical government ruled France until the First World War. It was a time of other endings, too. Europe's old social order, dominated by land-holding aristocrats, was drawing to a close, and a new order, based on entirely different values, was being born. No group reacted to these currents of change more directly and explored them more vigorously than those painters and poets and writers who lived on the Bohemian fringe of society.

THE COMMUNE

For Paris, a time of siege, rebellion and carnage

REPUBLIQUE FRANÇAISE.

VIVRE LIBRE, OU MOURIR!

LE 4 SEPTEMBRE 1870

GAILLARD

In 1870 there began in France a series of events that shook that nation to its roots. In July the French went to war with Prussia; six months later the fighting ended in smashing defeat, the tinsel empire of Napoleon III had collapsed, the capital was approaching famine and France was near anarchy. On March 26, 1871, the people of Paris seized the city and held it for 62 days. This was the famous Paris Commune, which conservative Frenchmen feared even more than Prussian militarism. The French government, aided by its enemy, Prussia, crushed the uprising. The Commune slogan was "Freedom or Death" *(above)*. The answer was death —on such a scale that the Seine ran red with blood.

THE START OF HOSTILITIES
Eager Foes, Apathetic Neighbors

Rarely has war been greeted with such enthusiasm as Paris and Berlin displayed for the Franco-Prussian conflict that warm, sunny summer of 1870. The plain people loved it. Parisians, upon learning on July 19 that funds had been voted for the war, filled the night air with roars of "*Vive la guerre!*" The war was to be the renaissance of France's fading glory. The newspaper *Le Figaro* opened a fund to present each *poilu* with a cigar and a glass of brandy, and a group of colorful Zouaves swashbuckled up the boulevards with a pet parrot that screeched incessantly, "To Berlin."

In Berlin the cry was "On to Paris!" The entire student body of Bonn University joined up, and crowds thronged the churches to pray for victory—and revenge for past humiliations at French hands.

Even more than the people, the leaders on both sides wanted the war: each saw it as serving his own ends. Prussian Premier Otto von Bismarck hoped it would bring the southern German states rushing to the side of Prussia—a glorious realization of his lifelong dream of a unified Germany under Berlin. To the weary Emperor Napoleon III, a successful war would restore the glitter of the imperial court, tarnished by military and diplomatic reverses.

Napoleon counted on the support of other nations against Prussia. But within weeks his expectations turned into the reality portrayed in the angry French cartoon shown here: while Prussia gobbled up Alsace and Lorraine, the rest of Europe—Britain, Austria, Italy, Spain, the Ottoman Empire—looked away. Only Russia paid attention. Intent on recovering losses incurred in the Crimean War, the Muscovites prepared to seize any advantage they could from the agonies of the Second Empire.

FLIGHT OF AN EMPRESS

Just 46 days after the war began, the Empress Eugénie stood rigidly in her opulent apartment in the Tuileries as the Minister of Interior read her an official telegram from her husband. "The army," it began, "is defeated." Eyes blazing, Eugénie screamed at her secretaries: "No! The Emperor has not capitulated! A Napoleon does not capitulate!" When a delegation of deputies asked her to abdicate, she flatly refused.

But the decision was not entirely in her hands. As word spread of the surrender, and the people of Paris rose up demanding Napoleon's removal, Eugénie's resolve crumpled and she fled from the mob. The composite photo at right—faked in accordance with a journalistic practice of the day—shows her quitting the Tuileries by a side door en route to exile in England. "I had no fear of death," she explained later. "All I dreaded was falling into the hands of viragos. I fancied them lifting my skirts, I heard ferocious laughter."

Meanwhile Napoleon's disaster had been played out on the battlefield of Sedan. In agony from a kidney stone "large as a pigeon's egg," his face rouged to conceal his pallor, the Emperor painfully mounted his horse on the battle's last day and rode through shellfire for five hours seeking death. In the end he was ingloriously taken prisoner, along with 104,000 men.

A REPUBLIC AGAIN

•

The French capital awakened September 4 to the newsboys' cry, "The Emperor a prisoner!" and Parisians erupted into the streets. A mob burst into the Chamber of Deputies and forced the legislators to agree to a republic—France's third. A list of names of prominent men was read aloud; those who won the loudest shouts from the crowd were elected. Of the new ministers shown at the left, the most important were General Louis Trochu, named President (top center) and Interior Minister Léon Gambetta (top right). Euphoria seized the city. With a vigorous new government, Parisians felt certain of ultimate victory.

A Prussian Thrust in the Night

On September 20, two German cavalry patrols linked up near St. Germain, completing the encirclement of Paris. A long stalemate followed. The Prussian army of 146,000 men (later increased to 236,000) was not strong enough to break into Paris; the French—100,000 army and navy regulars, an equal number of provincial recruits and 350,000 half-trained Paris National Guardsmen —were not strong enough to break out.

In the next four months there were hundreds of unsuccessful sorties such as this night attack by the Prussians on a Paris strong point. The city's defenses were substantial; a 33-foot wall ringed Paris, protected by a moat and 15 powerful forts. "The war," Bismarck complained to his wife, "is dragging out."

THE GREAT FAMINE OF NEWS

Next to food, news became the most sought-after commodity in besieged Paris. Nightly, Parisians of all kinds milled around the Interior Ministry (*right*), waiting for bulletins to be posted; this was one privation the rich shared with the poor. Twice Paris had no outside news reports for more than 20 days. One man beat the shortage. The Prussians permitted the U.S. Minister to bring through the lines a weekly copy of the *London Times*. But he had to agree to keep the contents to himself. The American was constantly begged to share the news; one Paris journal pleaded: "We gave you Lafayette . . . in return . . . we only ask for one copy of an English paper."

CANNON INSTEAD OF BONBONS

As the siege dragged on, Paris, to everyone's amazement, turned into one of the most powerfully armed fortresses in the history of Europe. "This modern Babylon celebrated for its . . . bonbons," wrote an incredulous Englishman from the capital, "now makes cannons, shots, shells and gunpowder by the ton." By September bullet production had reached 300,000 daily.

Parisian ingenuity became a weapon of war. Saltpeter for gunpowder was extracted from old plaster. Bronze and tin alloys replaced steel. With its factories de-

livering 21 cannon a day, the city soon boasted 3,000 big guns. The most famous was La Josephine, a huge, eight-and-three-quarter-ton bronze piece, shown above with its marine gunners at Fortress 40.

GAS BAGS TO THE RESCUE

Paris had a partial answer to the German ring of steel that isolated it from the outside world: hot gas. Soon after the siege began, the capital started sending men and messages aloft in cotton balloons filled with coal gas. The big gas bags were usually released at night, as scores watched. Of 65 balloons set free, 45 reached unoccupied France; others crashed, fell into enemy hands, or drifted onto foreign soil.

In all, the "Balloon Post" carried some 2.5 million letters—plus 164 humans, including Interior Minister Gambetta (below),

who flew out October 7 to organize the relief of Paris. Landing in southwest France, Gambetta raised new armies with incredible speed. Within a month a 100,000-man force routed a German army near Orléans, 70 miles from Paris, and occupied that city, giving France its first real victory of the war. When the word reached Paris, strangers kissed on the boulevards.

Gambetta now proposed that his troops attack from outside and link up with the besieged forces advancing from Paris. The proposal never had a chance. Gambetta was a leftist, President Trochu a monarchist; between them yawned an unbridgeable gulf. The two leaders proved unable to coordinate their moves—and the siege went on.

ALPHONSE DE NEUVILLE   EDOUARD DETAILLE

A New Height of Artistry

The siege settled into a routine—the daily drill of the National Guard, the nightly Prussian bombardment, Sunday strolls by citizens to inspect La Josephine and the fortifications. Artists Alphonse de Neuville and Edouard Détaille built themselves a tall platform (below) overlooking the siege and systematically sketched Paris' ordeal.

DINING ON SPANIEL AND ELEPHANT STEAK

When the siege began, Paris was crammed with edibles. Immediately after the Prussians invested the city, an official census reported 150,000 sheep, 24,000 cattle and 6,000 hogs—enough for two months. But the siege lasted twice that long. By December 8 a diarist noted: "People are talking only of what there is to eat." First came horsemeat ("equine sausage," it was called), previously eaten only by the poor. Shortly thereafter a reporter wrote, "I had a slice of spaniel the other day." On December 30, firing squads dispatched the beloved zoo elephants Castor and Pollux, one by one (*right*), and butchers started pushing choice elephant trunk at 40 francs a pound. Some Parisians even ate rats, drenched in rich sauces.

Afterward a scientist said that only six people died of starvation in the siege—but he added that another 4,800 infants and old people died of diseases "hastened by want of food."

FOR THE RICH, A RICH DIET

The siege sent some food prices soaring ninefold. The poor queued up at the municipal butcheries for as little as one and a half ounces of meat daily allowed them by the mayor's decree (right). But a wealthy Englishman said, "We get enough." Indeed, the rich did fare well, as this satirical cartoon shows. For them the booming black market made rationing a farce. Up to the very end one posh café offered roast beef, fowl, mutton, duck and two kinds of fish—for those who could pay the price.

A NEW REGIME, A DIFFERENT WAR

Just 193 days after the war began it suddenly paused. Bismarck and the French had struck a curious deal: France would lay down its arms for a brief period while it elected a new government. The new regime then would decide whether to sign a treaty or resume the war. After seven days of energetic campaigning, France elected 600 deputies.

The results stunned the Paris radicals. Only 200 of the new legislators were republicans; the rest were monarchists and conservatives. France had clearly stated its desire to end both the war and radical control of the capital.

The new chief of government was Adolphe Thiers, a right-wing historian (right). He named a likeminded cabinet that proceeded to fulfill its mandate.

GERMANS IN PARIS

Thiers and Bismarck argued for six days over the peace terms. At one point Thiers threatened to quit and let Bismarck govern France. Rather than take on that task, Bismarck yielded the point. In the end France gave up Alsace and Lorraine, plus a record one-billion-dollar indemnity. The German Emperor also was to be allowed to lead his troops into Paris in a victory march, and to occupy the city until the treaty was ratified. Going home Thiers wept. On March 1, 30,000 hand-picked German troops marched along the Champs-Elysées *(right)*.

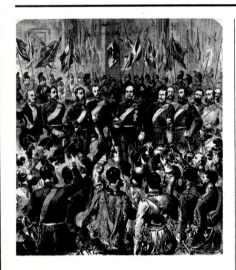

On January 18, in Versailles, Wilhelm was proclaimed the first German Emperor. He felt forced into the role by Bismarck and refused his Premier's hand.

A CITY BOWED BY DISHONOR

Paris blamed its defeat on General Trochu's military bungling. This bitter cartoon shows Trochu kneeling before the German Emperor, offering keys to the capital.

Threats were uttered in Paris against the conquerors but they were largely for effect. Some parading Uhlans lost their way and were set straight by obliging bystanders. Once Bismarck found himself in the middle of an angry crowd in the Place de la Concorde. He nonchalantly took out a cigar—and borrowed a light from the surliest looking man in the group.

The occupation lasted only two days before the peace treaty was ratified by the French. The Kaiser, grumbling at this disappointingly short stay, had to leave the

city with his troops. Parisians afterward scrubbed their streets to erase the shame.

A GRIM MEMENTO FOR PARISIANS

The departed Germans left this picture of themselves with their cannon. These weapons had hit Paris with over 300 shells nightly. The bombardment killed 97 Parisians, but the city took it in stride. During the siege a mother was heard to tell a child: "If you do not behave...I will not take you to see the bombardment."

REVOLUTION!

Paris vs. France: Birth of the Commune

The conservative government moved to control the radical capital. It named a rightist to command Paris' own left-wing National Guard, abolished the Guard's pay, and decreed that all debts not paid during the siege be made good within 48 hours. For many workers and shopkeepers this meant bankruptcy. Thiers then tried to seize 200 cannon from the Guard; the attempt failed and he fled to that home of monarchy, Versailles.

The capital, leaderless, held an election. On March 26 the new council voted to form the historic Commune. Now it was Paris against the rest of France. Marianne, symbol of the Republic, was shown on posters *(right)* holding a Commune flag inscribed: "Social Revolution, Equality, Justice."

THE MOB'S "JUSTICE"

Thiers feared to stay in Paris for good reason: two of his generals were slain there. General Claude Lecomte had been assigned to capture the National Guard's cannon, and actually had seized them when he discovered that he had forgotten the means to haul them away. While he waited for horses, an infuriated mob grabbed him, and also seized General Clément Thomas, who was out for an innocent stroll. The two officers were stood against a wall and shot—though not together, as shown in this composite photograph; Thomas was killed first.

A HISTORIC PRONOUNCEMENT

"The Commune is proclaimed!" With these words, Adolphe Assi, a mousy labor leader (below in Guard uniform), launched the Commune on March 28. Though Assi was a Marxist and the Commune was left-wing, it was inspired not by Marxism but by the 1793 Paris Commune that saved the Republic.

DISSENSION

The Commune seemed invulnerable. Thiers was in Versailles, and his only hope of returning to Paris—as the perceptive British ambassador pointed out—"appears to be that the members of the Commune will quarrel." They did, almost instantly. The Commune proved to be a hopelessly divided collection of do-gooders and grudge-holders and its sessions resounded with argument (above). The first meeting opened with a fight over who should preside, went on to a battle over capital punishment and adjourned in discord—a course it was to follow to the disastrous end.

A Strange Alliance

The Communards raged that Thiers and Foreign Minister Jules Favre—shown in Prussian *pickelhaube* helmets, firing a cannon at Paris—were German lackeys. That was not strictly true, but the two French officials had made an amazing deal with another well-known anti-Communard: Bismarck. To crush the Commune, Thiers needed part of the 400,000 French regulars held prisoners of war by Germany; Bismarck released them to Thiers because he feared that the Commune infection, unchecked, might spread to Germany's own socialists. This presented the spectacle of two countries, recently at war, joining to fight a group of rebels in one of them.

In only nine weeks the Commune's membership of 90 shopkeepers, workers and journalists produced 395 proclamations on subjects from art to toilets. The edict below abolished conscription and ordered all able-bodied citizens to enroll in the Guard. At the end, with danger pressing, the Commune spent hours debating whether to take religion out of the schools. The Commune's attention to trivia during a crisis, wrote one member, was "one of the most stupefying facts ever given to a historian to record."

RÉPUBLIQUE FRANCAISE

N° 42   LIBERTE—EGALITE FRATERNITE   N° 42

COMMUNE DE PARIS

La Commune de Paris DÉCRÈTE :

1° La conscription est abolie ;

2° Aucune force militaire, autre que la garde nationale, ne pourra être créée ou introduite dans Paris;

3° Tous les citoyens valides font partie de la garde nationale.

Hôtel-de-Ville, le 29 Mars 1871.   LA COMMUNE DE PARIS.

Marx's Mixed Emotions

Karl Marx, believed to have masterminded the Commune, actually was amazed at its establishment. Months before he had heaped scorn on his French followers who

wanted to set up a Paris Commune. When the Commune became a reality, however, Marx supported it. But he never stopped wondering why the Commune did not do something revolutionary—like grabbing the two million dollars in the Bank of France.

THE ILL-MANNED BARRICADES

The Commune was not oblivious to the trouble it was in. It built 100 barricades *(left)*, and ordered all males into the Guard. But the movement was losing appeal. Thousands of Parisians went into hiding rather than serve. The Guard, once 450,000 strong, now counted only one tenth that number.

Thiers also had troubles. Bismarck warned him to bring down "Red" Paris fast or Germany would do it. Thiers offered a Commune leader one million francs to open a gate; the bribe was refused.

A Monument in Ruins

With typical irresponsibility, the Commune's attention wandered even as the enemy stood at the very gates of the city. This moment of peril found the Commune devoting its energies to pulling down the 155-foot Vendôme Column, a bronze monument cast in 1810 of 1,200 captured Austrian and Russian cannon and topped by a splendid statue of Napoleon I in imperial regalia. On May 16, while 10,000 gaped and three bands played the "Marseillaise," the column tumbled in a heap of rubble (right). The operation required only a few hours—but it cost a fee of 6,000 francs for the demolition engineer.

"BLOODY WEEK" IN PARIS

On Sunday, May 21, Thiers's troops outside Paris were startled to see a handkerchief waving near the Point-du-Jour Gate. It was a Thiers sympathizer with incredible news: the gate was open; its exhausted guards had walked off. Within hours, 70,000 Versaillais soldiers poured into the city.

But now the Communards rallied, and fought with a singleness of purpose they had never shown before. Thiers's generals had estimated the battle would take three days; it took seven. The Versaillais smashed from street to street, using artillery point-blank. As the battle grew hopeless, the Communards crammed gunpowder, tar and petroleum into several buildings, including the Tuileries Palace *(left)* and touched off a wall of flame between the armies. Infuriated, the regulars dropped any lingering restraint and killed without mercy. The government had 873 losses, the city 3,000 —including scores of women and children.

Seven Days of Terror
A Contest of Atrocities

It took seven days to drown the Commune in blood. "I gave . . . orders," Thiers claimed later, "that the rage of the soldiers was to be contained." Instead it stormed unchecked. In Montmartre, where Thiers's two generals had died, the troops slew 49 Communards at random. And a diplomat saw the bodies of eight children, shot as alleged incendiaries.

The Communards countered with atrocities of their own. A petty official took 50 prisoners from La Roquette prison into a small courtyard *(right)* and, while a mob cheered, stood them up before a firing squad. When the shooting was finished, 51 corpses were counted. "Definitely one too many," the official said coolly. The Archbishop of Paris was also taken from La Roquette with four other

priests and shot. When the Versaillais found his unburied body, they promptly stood 147 Communards before a nearby wall and evened the score.

THE COMMUNE ON TRIAL

With the end of battle 38,000 Communards were in prison awaiting trial; among them were 650 boys under 17, jammed so tightly into stables and other makeshift prisons that some suffocated. For the next four years, 40 courts, such as the one in Blois shown here, ground out sentences; 25,000 were freed, 7,500 sent into exile and another 5,000 punished by fines or prison terms. Twenty-three were shot —a modest number compared to the earlier blood baths.

THE HEROINES OF PARIS

No one would soon forget the courage of the Communard women. They had "fought like devils, far better than the men," wrote an observer. Hundreds of them walked into captivity, some carrying babies. Cigar-smoking Hortense David had served on a Communard gunboat. Bravest of all was Louise Michel, part-time poetess, full-time revolutionary, the illegitimate daughter of an aristocrat. She had strode through battle, with a bayonetted rifle, screaming defiance. Her Women's Battalion, said one historian, was "about the only Communard detachment which showed spirit."

LOUISE MICHEL   HORTENSE DAVID

WOMEN BEHIND WIRE

After the second fall of Paris, the righteous women of Versailles came to look at the disreputable women of the Commune, who slept on the open quagmires of the Chantiers prison camp, awaiting judgment.

In this photo the notorious Louise Michel sits at the right of Hortense David (*left foreground*), who is shown, by fake photography, drinking from a whiskey bottle. Convicted of setting fire to buildings, Mlle. Michel asked to die with her friends, but the court refused and sentenced her to life imprisonment in a penal colony on Nouméa Island in the Pacific. When she was finally freed in the amnesty of 1880, she went straight back to revolutionary activity in France and drew three more sentences before going into exile in England. There, a little old lady in black, she died, exulting, on hearing of the 1905 Russian Revolution.

The Terrible Price

The bloodletting was over, but the memory of bodies strewn in the streets was seared into the nation's soul. In all, almost 25,000 people had been killed—10 times as many as had died in the 15-month-long Reign of Terror of the Revolution of 1789. France would never forget those few days, the most violent in the nation's history.

7

ARTISTS AT
THE CROSSROADS

The Age of Progress would one day be considered as remarkable for its art and literature as for its commerce and industry. It saw the ebb and flow of a great many movements—movements that multiplied and obsolesced as rapidly as machinery; movements that converged and conflicted as frequently as theories of science, politics and religion. Among them the foundations of modern art and modern literature were laid.

The visitor to the Great Exhibition in London in 1851 could have had little inkling of this. Although large and popular exhibits were devoted to the fine arts, in inventiveness they could not equal the engineering of the roaring machinery or the modern design of the Crystal Palace's glass shell. The Great Exhibition was a celebration of the useful and practical, a display of the technological triumphs of an acquisitive society. The exhibits devoted to art at the Crystal Palace were more of an index of middle-class taste and sentiment than they were of artistic achievement.

Little painting was shown, but there were large displays of overelaborate furniture and Neoclassical sculpture. Two of the sculptures were especially popular. One was an allegorical group called *The Lion in Love*, by Guillaume Geefs of Belgium; the other was *The Greek Slave*, by Hiram Powers of the United States. The Belgian entry showed a naked young woman of Classical Greek appearance seated on the back of a docile lion. "The monarch of the forest," a critic explained, "unable to resist the seducing loveliness of a nude female who is . . . fascinating him with her eyes, is quietly submitting to be deprived of his claws."

Hiram Powers' marble slave girl was also Classical, and naked except for her chains. A critic cited the figure as exhibiting "an extraordinary refinement of imitation"—and "imitation" is the key word.

The early half of the 19th Century had seen revivals of Classical and Gothic art forms; by the middle of the century revival had degenerated into imitation as the commercial uses of art were exploited. The artists who were the most successful were those who catered to the wish for ornamentation and sentimentality. One of the most popular and probably the richest—when he died in 1873 he left an estate of £300,000—was Sir Edwin Landseer, a man of technical talent but little disposition for innovation who painted his way from country house to country house, producing one portrait after another of well-bred dogs and children.

Landseer had no illusions about his work. "If people only knew as much about painting as I do," he once said, "they would never buy my pictures." But his pictures sold as fast as he could paint them. There were artists like him in every country of Europe. When they did not paint portraits, stags and landscapes, they painted Classical nudes or nymphs and Renaissance angels, sometimes combining lascivious suggestion with moral sermonizing, often telling stories drawn from history, religion and ancient mythology.

The success of such works reveals some of the criteria that guided the average person's taste in art around the middle of the 19th Century. He generally wished to be told a story, to be diverted and to receive a little moral instruction. The art that satisfied these demands relied on the imitation of a remote past. It did not deal with the life that people saw around them, nor delve into social problems; it did not demand that the viewer inquire into the unfamiliar; it conveyed no universal message, and it encouraged sentimentality.

Part of the blame for the irrelevance of so much mid-century art rests with the arbiters of public taste—the critics and the directors of such institutions as the Royal Academy of Art in London and the Ecole des Beaux-Arts in Paris. These men were enamored of the past no less than the public, and they guided themselves by rigid rules of "correct" drawing and "proper" subject matter.

They also held the fate of painters in their hands. Every year the Salon des Beaux-Arts held an exhibition that was a national event and something of a state occasion. For a painter to be accepted for showing was essential to his professional life; rejection meant that his pictures would be

returned to him bearing an *R* for *Refusé*, a stigma in indelible ink on the canvas that often rendered the hapless painter unable to sell or exhibit his picture afterward. All painters therefore, even those who professed to scorn it, coveted Salon approval. When they did not get it, poverty and derision were their usual rewards. Jean-François Millet, whose paintings of peasants were denounced because they "reduced human beings to clods," once sold six drawings to buy a pair of shoes; those drawings may be worth upward of $20,000 today. Often Claude Monet could not work because he was unable to afford the price of paints, and Auguste Renoir, an artist of singularly cheerful disposition, wrote to a friend, "We don't eat every day."

It might be expected that little of promise would flourish in such an atmosphere. Yet despite the hold of the Salon, despite the exigencies of commercial success, despite the prevalence of sentimental portraiture and moral storytelling, the latter half of the 19th Century was in fact a time of great ferment. There were men in the world of art who were stirred no less than their contemporaries in science and industry by the spirit of innovation that animated the Age of Progress. These men were destined not only to loosen the grip of the academies, but also to revolutionize theories of subject matter, composition and color.

The first significant new movement in art was Realism, which grew in France in the 1850s. The term "Realism" did not mean portrayal with photographic accuracy; it meant the presentation of contemporary facts of life, usually calling attention to the unfortunates of society and never glossing over the unpleasant. The first painter to accept the term and apply it to himself was Gustave Courbet, an earthy man from Ornans in eastern France. Courbet came to believe that the prevailing romanticism in art was only an escape from the harsh

realities of the time, and that the artist must portray his own experience. "Show me an angel," he said, "and I will paint one."

Since no one showed him an angel, Courbet painted large and vibrant canvases portraying ordinary French life. One of his first such paintings was *The Burial at Ornans*, which shows a group of grieving peasants around a grave. Such genre painting, showing everyday scenes, was not new, but painters generally sentimentalized the peasants and made them seem happy in their humdrum lives. Here Courbet was presenting them in epic proportions. The canvas, to begin with, measures about 10 by 22 feet—a size generally reserved for portrayals of historic events. Courbet was suggesting that the suffering of peasants was equally deserving of human concern. He later referred to the painting among his friends as "The Burial of Romanticism." The critics of his day were not so cheerful. One of them said of it, "Probably never has the cult of ugliness been practiced more frankly."

Courbet was undismayed by the criticism of the press and unmoved by pressure from the Salon. When Count Niewekerke, who was Superintendent of Fine Arts in Paris and a powerful figure, once suggested to him that he try to put more charm in his pictures, Courbet indignantly refused to do so. And when the Salon des Beaux-Arts rejected a group of his paintings in 1855, he stubbornly hired

a shed nearby, named it the "Pavilion of Realism," and showed his paintings alone. Hardly anyone paid the one-franc admission to see what he had to show. The artist Delacroix, who did, reported that he was there alone for nearly an hour. "I discovered a masterpiece," he said. "They have refused there one of the extraordinary works of our time."

Delacroix was referring to *The Painter's Studio*, a canvas that shows Courbet at work among more than 20 people. To the right of the painter is a group representing the well-to-do and the intellectuals of Paris—art patrons, a critic and a novelist. To the left are the destitute and the exploited of society—a Jew, a prostitute and a thief. Except for a nude model who peers over Courbet's shoulder, and a child who looks up from his knee, no one seems to take any notice of what is going on—an indication of the individual's alienation from society, and an early commentary on social stratification.

By the end of the 1850s Courbet was not unique in espousing Realism, nor was he alone subjected to abuse. A host of other painters were doing the same, and with similar consequences. The more they persisted, the more stubborn became the academic resistance to them. In 1863 the Salon des Beaux-Arts turned down 3,000 out of the 5,000 works submitted for the annual exhibition—the highest proportion of rejections anyone could remember. Rumors began to circulate in the press that the jury had committed a "massacre."

Alert to the public clamor, the Emperor Napoleon III paid an incognito visit to the Salon one afternoon to inspect some of the paintings that had been refused. Finding many of them "quite as good as those accepted," he summoned Count Niewekerke to ask for an explanation. Niewekerke replied that the number of would-be artists had increased in past years, and that it was essential to "dam this flood of individualities."

Napoleon disagreed. Gossips said he wished to

spite Niewekerke, a pompous man and the lover of Napoleon's niece; in any event, Napoleon always heeded public opinion. A few days later he announced that he would hold a *Salon des Refusés*, to let the public see the discarded paintings.

The Salon des Refusés showed among its several paintings the works of Edouard Manet, Camille Pissarro, Paul Cézanne and dozens of others who have since become famous. For several weeks three or four thousand visitors flocked there every Sunday—drawn, no doubt, by the notoriety given the show by Napoleon's intervention and by the conflicting critical evaluation, which ranged from warm praise to exclamations that it was "hilarious."

The most sensational picture of the exhibit was Manet's now-famous *Déjeuner sur l'herbe*, which shows a nude woman picnicking with two frock-coated bourgeois men. When the royal family visited the exhibition, the Empress averted her eyes, pretending not to see it; Napoleon, less dainty, intoned: "This painting is an offense against modesty." Public and critics followed his lead and condemned the picture on moral grounds—giving the Beaux-Arts jury the chance to say in effect, "We told you so."

Nobody noticed that the three major figures in the picture were inspired by a Raphael design of three river gods—an inspiration that 19th Century Neoclassicists would have found unexceptional. Even less did people perceive the satire on bourgeois convention. What Manet had done was take a Classical pose, put it in a contemporary setting, and add a fillip of fantasy. The result was neither Classicism nor Romanticism, nor the new Realism of everyday experience; it was a declaration of artistic freedom.

Baudelaire offered Manet the thought that "ridicule, insult, injustice, are excellent things," and the painter Degas assured him that the criticism had made him as famous as Garibaldi. But unlike the irrepressible Courbet and the cheerful Renoir, Manet was a brooding man and profoundly distressed by the poor reception given his works.

Two years later, in 1865, Manet had no more luck with the reception of his *Olympia*, which shows a nude reclining on a couch in a somewhat Classical pose. It was not the nudity per se that incensed Manet's critics; nudes were all right if, like Venus, they emerged pristine from the sea. Manet's *Olympia* shocked with her impudent stare, with her partial adornment (she wears a ribbon around her neck) and because she lies on rumpled bedsheets. "What is this odalisque with a yellow belly, a degraded model picked up I don't know where?" asked one irate critic. "What's to be said . . . for the black cat which leaves its dirty footprints on the bed?" demanded another. Even Courbet, whose own subjection to abuse had evidently not disposed him kindly toward others' innovations, said that *Olympia* was as flat as a playing card.

Courbet was referring to Manet's style, for here, as in subject matter, Manet was departing from tradition. He was abandoning chiaroscuro (modeling with a full range of lights and darks), and substituting dimensionless backgrounds for perspective. Chiaroscuro and perspective were two techniques developed during the Renaissance and honored by Western painters ever since; their abandonment by Manet was revolutionary. He developed the technique further, and other painters followed.

In 1874 a new group of painters who for years had been consistently rejected by the Salon assembled their work for exhibition in a studio on the Boulevard des Capucines. Among them were Claude Monet, Auguste Renoir, Camille Pissarro and Alfred Sisley, some of whom had exhibited at the Salon des Refusés 11 years before. The title of a painting by Monet, *Impression: Sunrise*, led a hostile critic to call him and his fellows "Impressionists"—a term that Monet never ceased to resent, though it was soon to lose its pejorative connotation. He preferred the word "instantaneity," for his main concern was conveying the fleeting and the spontaneous, the changing effects of light and its perception by the eye at a given moment. This he and his confreres achieved with short brush strokes that broke up a field of color into a profusion of hues.

Like Courbet and Manet, Monet and the others of his group were influenced by Realism, but they were Realists of a new breed. Like Courbet, they took their subject matter from ordinary life, but they rid their painting of moral nuances and sociological significance and made it light-hearted instead. Like Manet, they discarded perspective and detail, but they also took the new technique out of doors, to bathing parties and picnics, or into the bourgeois world of dance halls, cafés and the theater. They observed the effects of light in a quasi-scientific fashion, for by the 1870s science had taken hold of the popular mind.

None grew more determined to seek the varying effects of light than Monet. In his search for "instantaneity" he would set out for the fields surrounding Etretat, near Le Havre, accompanied by a troop of children who helped him carry a load of canvases. The writer Guy de Maupassant often watched him, and reported: "He took the canvases up and put them aside in turn, according to the changes in the sky. Before his subject, the painter lay in wait for the sun and shadows, capturing in a few brush strokes the ray that fell or the cloud that passed." Monet once painted 15 different versions of the same haystack at different hours of the day. So obsessed was he with catching the transitory that, to his own consternation, he found himself studying the changing tones of his young wife's face as she died.

If Realism took over painting in the third quarter of the 19th Century, it found an even larger expression in literature—most especially in the nov-el, which was perhaps the most popular art form of the era. And in literature as in painting, Realism was inspired first by an expanding social consciousness and second by science.

The novel had from its birth been a blend of contemporaneous Realism and Romantic diversion; with the 1850s came a strong tide of moral judgment. In England Charles Dickens, who had been put to work in a warehouse at the age of 12, wrote of the sorry lot of the child in the factory and of the brutality that breeds in slums. For his sentimentality he could be called a Romanticist; but his bleak picture of life among the poor makes him a Realist. William Makepeace Thackeray wrote astringent satires of middle- and upper-class pretensions and of the vanity of life.

In France, Gustave Flaubert wrote even more pitilessly of bourgeois life on the Continent. His most renowned work, *Madame Bovary*, which appeared in 1857, was at first glance the story of a flighty provincial housewife and her escapades. More subtly, it was a cruel diagnosis of the petty society that stifled her. On its publication, Flaubert was sued by the government for having committed an "outrage to public morals and religion." He was acquitted, but his angry middle-class readers had correctly perceived that he had attacked the very basis of their comfortable society.

The Realists were undoubtedly idealists at heart; they railed against the injustices of society and longed to better it. But this was an age in which romantic illusion was daily yielding to scientific detachment. No sooner had Realism gained a foothold in literature than it was overtaken by Naturalism, a movement in which writers aimed at describing the world with the objective precision of scientists. Naturalist writers, like Impressionist painters, called for a careful observation of life without guidance from morality or tradition. But whereas the painters were concerned with the physics of light, the writers were concerned with contemporary biology. Thus they assigned to heredity and environment the roles that once the gods had played in determining the fate of a hero.

No one took Naturalism more seriously than Emile Zola. Comparing himself sometimes to a surgeon who subjected the human body to clinical analysis, sometimes to a police commissioner who examined the facts of a crime, Zola scrutinized all aspects of contemporary society; even the most sordid was not beneath his examination and portrayal. When writing a novel he studied factual monographs and prepared dossiers for his characters, as if they were live people. Between 1871 and 1893 he wrote 20 novels, a series together entitled *Les Rougon-Macquart,* which told the story of different branches of the same family, who had hereditary traits in common but who developed differently according to their social environments.

Labels come readily to the founders of a trend, even to those who carry it to its conclusion. Not so easy is the classification of genius, the designation of men who take the techniques of their predecessors and create an art that defies definition, that belies all formula, that transcends even the laws of logic. Dickens, Thackeray, Flaubert and Zola were masters of their craft, talented, imaginative and inventive. But the culmination of the effects on art of 19th Century forces was to come from other men—indeed, it was to come from Russia, which lay beyond the Europe of Progress.

Russia had come under the influence of Western literature just as she had felt the effects of Western industrialism; in the early 19th Century, for example, her writers often imitated the Romantic poets Byron and Scott. But about 1860 Western Europe was astonished to find herself outclassed by Russia with novels that not only portrayed the shifting social scene, that not only took heredity and environment into account, but that also plumbed the depths of the human spirit as had not been done since Shakespeare.

The first of the Russian giants was Ivan Turgenev, who had traveled in the West and subsequently settled there. In 1852 he published *A Sportsman's Sketches,* essays recounting his walks through the woods and fields of his native Bolkhov, where he conversed with the local peasants. His poetic accounts of their lively imagination and unaffected dignity presented them, like the people in Courbet's *Burial at Ornans,* in a startlingly human light. Because it credited them with surprising intelligence and wisdom, the book had the explosive effect that Harriet Beecher Stowe's *Uncle Tom's Cabin* had in the United States, almost at the same time. Less than a decade later, when Lincoln was preparing to emancipate the slaves in the American South, Czar Alexander II, who is said to have read Turgenev's book, had already freed the serfs in Russia. Here was dramatic evidence of what the Roman poet Horace had meant when he called poets the first instructors, and what Shelley, more recently, had meant when he called them "the unacknowledged legislators of the world."

The second great Russian writer was Leo Tolstoy, a nobleman who shared Turgenev's compassion for the serfs; he twice made unsuccessful attempts to establish schools for them and at the end of his life put aside his title to live among them. Of his two most famous novels, one, *War and Peace* (regarded by many as the greatest novel ever written), is a panorama combining history and contemporary life. It has 60 heroes and some 200 lesser characters drawn from the whole spectrum of society—court figures and common people, generals and lowly soldiers. His other famous novel, *Anna Karenina,* combines depth of character study with a penetrating analysis of social convention. Tolstoy, who deplored man-made government, social structure and dogmatic creeds, wrote with unprecedented insight into the effects those institutions have on human character.

Tolstoy vies for honors as the world's greatest novelist with Feodor Dostoevsky—an epileptic, an intellectual revolutionary and a gambler who lost most of his earnings betting. Preoccupied with human crime and guilt, and believing that even the vilest of human beings was capable of redemption, Dostoevsky anticipated the lessons of modern psychoanalysis by nearly half a century.

What the Russians Turgenev, Tolstoy and Dostoevsky did to raise fiction to a new height, a Norwegian, Henrik Ibsen, did for the drama. Ibsen, an introvert with a lifelong grievance against hypocrisy and cant, was intimately familiar with the sham of "respectable" society. He had lived in comfortable circumstances until the age of eight, when his father was bankrupted and the impoverished family was ostracized by the town in which they lived. Discarding such Romantic conventions as dueling and lovers' trysts, dramatizing instead the tensions that lie behind family dissension, and daring to speak the socially unspeakable, Ibsen brought contemporary life to the stage. Like the Russian novelists, he portrayed universal human passions that know no national boundaries.

By the last third of the 19th Century many writers and other intellectuals could see nothing but ever-increasing ugliness in European society. Many of them, wishing to retreat from the brutal realities of industrial civilization, reacted against the explicitness of Realism and the harshness of Naturalism, and turned for solace to a subjective world of shadowy images evoked by symbols.

Symbolism began with poetry, but it soon spread to the novel, to painting, and even to music. The Symbolist writers wished to capture "a sense of the ineffable"; their purpose was not to depict life or to reproduce reality, but rather to suggest the world of the spirit through imagery, and through that to

THE MANY SCHOOLS OF ART

REALISM, one of the first new styles to evolve during the Age of Progress, was an attempt by Courbet and others to depict the world exactly as it existed. It concentrated on the lower classes and the commonplace.

IMPRESSIONISM used colors broken into component hues to represent light, illuminating scenes as they appeared at an isolated moment. Among its most celebrated practitioners were Manet, Monet, Pissarro and Sisley.

POST-IMPRESSIONISM is a term loosely applied to several schools that followed Impressionism, but it usually refers to four men who composed forcefully, using intense colors: Van Gogh, Gauguin, Seurat and Cézanne.

EXPRESSIONISM, which flourished in the early 20th Century, was primarily German. Its members, notably Kirchner, used strident colors and distorted forms in an effort to express their personal vision of the world.

FAUVISM, literally the art of "the Wild Beasts," was based on the belief that color and spontaneous expression should dominate a painting. Among its famous names were Matisse, Vlaminck, Rouault and Dufy.

CUBISM, a reaction against Fauvism, concentrated on shapes, breaking them down into cubes and planes until the object portrayed became a series of geometric surfaces. The movement was led by Picasso and Braque.

FUTURISM attempted to capture motion on canvas in order to express the dynamic modern world of racing motor cars and crowded city streets. It was practiced by such Italian artists as Boccioni and Severini.

ART NOUVEAU, an ornate style of swirling, natural forms, was employed mainly in decorative and commercial work, but also in painting and fine prints. Its more renowned exponents included Klimt, Crane and Beardsley.

achieve "total experience." Believing that art could transcend life, they sought the exotic and the unknown, and the further they carried their idea, the more difficult it became for the average person to understand them.

The leader of the Symbolist poets was the Frenchman Stéphane Mallarmé, who held Tuesday afternoon receptions at his flat on the Rue de Rome in Paris, where he expounded his views for the edification of aspiring young Symbolist poets. Mallarmé deliberately made his verse suggestive by omissions, peculiar syntax and unconventional diction. "To name an object," he said, "is to destroy three quarters of the enjoyment of a poem, which is made up of the pleasure of guessing little by little." Art had come a long way from the conventions of 25 years before, when stories—and paintings—were explicit and left little to the imagination.

In the German composer Richard Wagner the Symbolists found a hero. In 1885 several Symbolist poets, including Mallarmé, joined to contribute to a new magazine, the *Révue Wagnerienne*, which was founded to publicize Wagner not only as a musician, but also as a poet (he wrote his own libretti), a great thinker and the creator of a new art form.

As developed by the Italians, opera was essentially a play embellished with music; Wagner made the music an integral part of the production. Moreover, he took his plots from Teutonic mythology, a subject sufficiently remote from contemporary civilization to delight the Symbolist poets. He shared the Symbolists' view that art should provide a "total experience"; in fact, his choice of opera as his medium sprang from his belief that through it all aspects of art—drama, dancing, painting, architecture, orchestration—could be brought together in a single vast synthesis, which he called *Gesamtkunstwerk*, or "total work of art." Through opera and the Teutonic mythology on which he based it, Wagner set forth his views of man, nature and God.

Similar trends were underway in painting. In the 1880s, Impressionism was superseded by Post-Impressionism, which in turn was to divide into two streams. One stressed the evocation of emotion; the other stressed the disciplined application of the intellect to structure and composition.

The Post-Impressionists were not a school; they were a diverse group united only in their abandonment of the thought that art should reproduce reality as closely as possible. They did not wish to overthrow Impressionism, but rather to develop it further. They were not content to paint transitory appearances, but rather sought the enduring; instead of portraying the light-hearted, they sought to understand man and life. Of the two streams of Post-Impressionism, the first is best illustrated by Paul Gauguin and Vincent van Gogh.

Gauguin was the archetype of the modern bohemian, the nonconformist who flouts convention and lives in isolation from society. By birth he was a member of the solid French middle class, by profession a prosperous stockbroker, and by avocation an art collector and occasional painter. At the age of 35, finding himself oppressed by the race for material gain, he gave up his career, deserted his wife and four children and went north to paint in Brittany, where he lived a hand-to-mouth existence. Touched by the simple faith of the native Bretons, he painted *The Vision after the Sermon*, a picture that shows the Biblical Jacob wrestling with the Angel, while peasant women in Breton dress look on with reverence. In style the picture was a departure from Impressionism because it used fields of solid, rather than broken, color; in subject matter it was one of the first applications of modern art to a Biblical theme.

Some years later Gauguin retreated even farther, to Tahiti. Colonists were now streaming to Asia and Africa to impose their material way of life on primitive peoples, and missionaries were carrying the Christian Gospel there, but Gauguin went to learn rather than to preach. In an age when most of his contemporaries were pursuing wealth and the rewards of advancing technology, Gauguin lived on boiled rice, dry bread and mangoes, went as naked as the natives and learned the local language, lore and mythology. In Tahiti he painted his masterpiece *Whence Do We Come? What Are We? Where Are We Going?* The picture shows a group of natives in childhood, maturity and old age, and is an attempt to find, through the symbols of a primitive culture, universal truths of which industrialized man has lost sight. Gauguin did not attempt to supply the answers to life; he merely posed the questions.

Another bohemian was Vincent van Gogh, a small, ugly and charmless man who yearned for a niche in society but could not find one. He came of a clergyman's family, entered a theological seminary but failed to graduate, then went to work as a lay preacher among the coal miners of the Borinage in Belgium; but he was so intense and so awkward that the miners and their children laughed him out of town. Trying once more to make a career of religion, failing again, and twice spurned in love as well, he turned at the age of 27 to painting. Knowing that he was a faulty draftsman, he entered school with youngsters who regarded him as the class freak and ridiculed the zeal with which he applied himself to learning.

But at last Van Gogh succeeded. Using brilliant colors, and distorting forms to suit himself, he conveyed on canvas his mystic faith in a creative force that he believed animated all forms of life. In the process, he found a release for his own restless emotion; a radiant sun in the sky shows an ebullient mood and the gnarled trunk of an olive tree conveys his pervasive sadness. In 1888 he began suffering epileptoid seizures; despairing of a cure, he committed suicide. The man who had failed at almost everything else he tried in life had, in the space of 10 years, succeeded in making of art an instrument of intense personal expression.

Two other Post-Impressionists carried Impressionism in a different direction. They were Georges Seurat and Paul Cézanne, who thought Impressionism too formless and lacking in discipline.

Seurat was the antithesis of the bohemian escapist; he found inspiration in industrial society. Fascinated with science, he enthusiastically applied its findings to his painting, learning from physicists' studies of optical effects and calculating mixtures of color accordingly. But to this new knowledge he added the Classicists' determination to arrange the figures of his paintings according to strict laws of composition.

Like the Impressionists, Seurat painted scenes from contemporary life, but his figures, which are reduced almost to silhouettes, appear immobile instead of in quick motion. His most famous painting, *A Sunday Afternoon on the Island of La Grande Jatte*, is a large canvas measuring about 7 by 10 feet. It consists of thousands of tiny dots applied with painstaking precision—a feat that could be accomplished only by a man with a taste for the difficult; it took him over a year to do. This technique, which is called Pointillism, renders its effect on the eye according to the same principle that governs modern half-tone engraving. A photoengraving is a mass of infinitesimal dots that merge before the naked eye to create different tones of gray; in Seurat's painting dots in such colors as red and blue, or blue and yellow, coalesce to form masses of purple or green.

Of all the Post-Impressionists, perhaps the most relentless in his search for a new form was Paul Cézanne, a surly, rough-mannered man who was born in 1839 in Aix-en-Provence, where his father was a banker. His closest boyhood friend was Emile Zola, who persuaded him to give up law for painting, and who later achieved fame at a time when

Cézanne was still carting his paintings in a wheelbarrow to the back door of the Salon des Beaux-Arts, hoping in vain to have some of them accepted. When Zola published his novel L'Oeuvre—the story of a painter who defied popular taste, and in whom Cézanne was recognizable—Cézanne was bitterly hurt, for it seemed as though his friend had made a travesty of his struggles. He never spoke to Zola again.

Cézanne's career spanned most of the second half of the 19th Century and his style accordingly shifted from one school to another. In his youth, in the 1860s, he was absorbed in Romanticism; in the next decade he showed his pictures with the Impressionists. In the 1880s he turned away from Impressionism, saying, "I do not want to reproduce nature, I want to re-create it."

In his effort to "re-create" nature Cézanne was the most revolutionary painter since Giotto 600 years before. Fascinated with Classical composition and perspective, he sought, like Seurat, to impose them on Impressionism; but in place of the Classical use of line to convey them, Cézanne used different colors in juxtaposition.

Cézanne applied this method to portraits, still lifes, and most especially to landscapes. He became obsessed with the problem of conveying the shape of Mont Sainte-Victoire, a craggy mountain peak near Aix-en-Provence, where he lived, and painted many versions of it. In none of them is there any hint of human life, for Cézanne never shared the Impressionists' interest in "slice-of-life" scenes, or in movement and change. "I want to make of Impressionism," he said, "something solid and durable like the art of the museums." In this Cézanne succeeded; he ranks today as one of the greatest painters of the last hundred years. He died in 1906 of pneumonia, which he caught while painting in a storm—fulfilling a wish he had earlier expressed, that he wanted to die painting.

The decade that followed Cézanne's death saw a proliferation of new movements: Expressionism, Cubism, Fauvism, Futurism and several others. Their fleeting appearances correspond with the acceleration of all social phenomena in the final decades of the Age of Progress. The artists were conveying what philosophers, scientists and statesmen were demonstrating in a variety of other ways: that an old order was fast disappearing and that man, in undergoing change, was in a state of confusion.

During the same period, James Joyce was at work on Ulysses, a novel of interior vision, a development of the Symbolist idea; it was to be the most revolutionary piece of fiction in a century. The poet Guillaume Apollinaire was experimenting with poetry written in the manner of the Cubist painters, arranging the words in geometric patterns on the page and eliminating punctuation. It became as difficult to find the meaning in an Apollinaire poem as to seek a story in a Cubist painting.

Writers and painters were marching away from the traditional just as surely as scientists and industrialists had done. Ironically, their advances did not meet with the approbation given science and technology. In ways not easily recognized by their contemporaries—or perhaps too difficult to accept—they were describing the essence of their age: the coexistence of diverse ideas; the isolation of the individual from society; discord under a peaceful surface. Like the bourgeoisie, the complacent pursuers of materialism whom many of them scorned, the intellectuals were optimistic about the future; they believed that a better world lay just around the corner, and that the making of it was in their hands. But consciously or not, they were expressing a conflict that underlay the peaceful surface of European society at the turn of the century—a conflict that was to erupt with the most fearful consequences that history had ever known.

THE UNDULATING LINES *of a bronze dancer and a floral wallpaper design exemplify the Art Nouveau style.*

A FLAMBOYANT ART

During the 19th Century, the decorative arts languished in a stagnant preoccupation with Classicism until an exotic new style burst upon the scene in the 1890s. Hailed as "Art Nouveau" (and nicknamed the "tapeworm style"), it was an attempt ·to create a fresh beauty by combining forms found in nature with meticulous hand craftsmanship. Soon its plant and animal motifs and sinuous, flowing lines had swept the world. Art Nouveau was found everywhere: in furniture, jewelry, posters, books—even in entire houses built and decorated in the new style. The movement disappeared as quickly as it had come—it was outmoded by 1914—but while it lasted, it was a perfect expression of its extravagant age.

SETTINGS FOR
THE SALON SET

One of the first footholds of Art Nouveau was interior decoration; furniture, lamps, wallpaper, fabrics, even plumbing fixtures were designed in the new style. In 1895 the movement received its name and what amounted to a formal introduction when an art dealer, Siegfried Bing, opened a shop in Paris called La Maison de l'Art Nouveau.

Here Bing displayed pieces like those shown in the room at right, including glass lamps by the American designer Louis Comfort Tiffany, intricately curving furniture, ornate wallpaper, pillows and art objects. Bing's wares were almost prohibitively expensive; all were handmade and lavishly ornamented. But they appealed to wealthy Parisian socialites for whom the height of fashion was the *salon*—a glittering gathering of artists, writers, wits and dandies invited for an evening of urbane conversation in their hostess' Art Nouveau drawing room.

As the vogue spread, ladies wore dresses by Art Nouveau designers so that their persons as well as their surroundings would have the flourish of the new style. Bing and his artists stressed the idea of "total design"—having a single artist create every object that went into a house. One Belgian architect carried this concept to its logical extreme. He invited the painter Toulouse-Lautrec to luncheon, and chose even the color and texture of the food to blend with the Art Nouveau décor.

THE CULT OF CRAFTSMANSHIP

Art, the early proponents of Art Nouveau firmly believed, should be a workman's expression of his pleasure in work. Nowhere was this belief more faithfully carried out than in Art Nouveau glassware, particularly that of the French master, Emile Gallé, who designed the lamp and several of the vases in this photograph. Gallé employed over 300 men in his shop in Nancy, but each piece that came out of it reflected his taste and style and often included his own personal workmanship. Frequently he left traces of tools he used, such as the scratch of a carving wheel, or purposely included flaws like air bubbles in the glass to insure the individuality of each piece.

Not all Art Nouveau craftsmen, however, shared Gallé's taste for graceful elegance, and some catered to an esoteric fancy for the grotesque, or what they called "decadent." For example, the ceramic frog vase in the left foreground, while executed with meticulous care, also intentionally induces a slight feeling of disgust.

A FLAIR FOR SELF-ADORNMENT

Nothing pleased Art Nouveau craftsmen more than creating exotic objects to display their virtuosity. In jewelry they indulged this taste to the fullest extent, designing exquisite baubles for a theatrical society which performed as if it were constantly on stage. When Parisian ladies, in their splendor of feathers, ruffles and lace, went to the races at Longchamp or strolled in the Bois de Boulogne, they adorned themselves with jeweled hairpins, hatpins, necklaces, rings, bracelets and brooches, many of them in Art Nouveau style. Even the men went in for extravagant stickpins like those shown at the right; one fashionable patron of Art Nouveau went so far as to attend an afternoon concert clad entirely in mauve—with a bunch of violets fastened at his neck in place of a tie.

The master jeweler of Art Nouveau was the Frenchman René Lalique, who designed the dragonfly chimera above. One of his best patrons was the actress Sarah Bernhardt, whom Parisians and Londoners worshiped as "La Divine Sarah" for some 40 years. With her extravagant taste and insatiable desire to shock people, she personified the opulent age by living in an aura of scandal and publicity. She was enchanted by Lalique's jewelry, and when she bought it, it quickly became the rage.

A FANTASTIC DRAGONFLY—*complete with a green female torso, gold griffon's claws, and wings inlaid with enamel and semiprecious stones—was designed by Lalique to be worn as a brooch.*

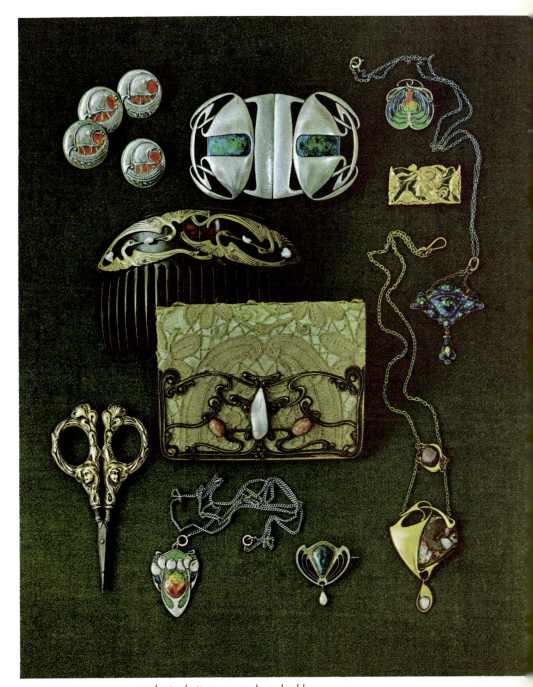

AN ARRAY OF TRINKETS—*pendants, buttons, a comb, a buckle, sewing scissors, a case for calling cards—display the Art Nouveau craftsman's love of elaborate designs for commonplace objects.*

JEWELED STICKPINS, *held in a Tiffany vase (left), were worn by fashionable European dandies at the turn of the century. The bumblebee pin (top) was used to attach a woman's hat to her hair.*

A COLLECTION OF VOLUMES *bound and illustrated in the new style includes two German works for children, Virgil's "Bucolics," a literary periodical called*

The Yellow Book'' and a biological journal containing studies of octopuses.

ART NOUVEAU TYPE *emphasized style more than legibility.*

BOOKS DESIGNED
TO PLEASE THE EYE

From its inception, Art Nouveau appealed to writers and artists like Oscar Wilde and Aubrey Beardsley; as a result, the style flourished in the field of book and magazine design. Every kind of publication, from rare limited editions of classics like Hugo's *Notre Dame de Paris* to inexpensive children's books, was lavishly illustrated and bound by craftsmen working in the new style. These artists particularly relished books with marine settings, because waves and tentacled sea creatures were perfectly suited to Art Nouveau's curling lines. They even designed special type faces with characteristic swirls and flourishes. Although he was one of its patrons, Oscar Wilde mocked the excesses of the craze: in his novel *The Picture of Dorian Gray*, Wilde had his effete hero order his favorite book bound in nine different ways, to suit any conceivable mood he might feel when selecting a copy to read.

A FLOWERING OF
POSTER ART

One Art Nouveau medium which was not limited to the eyes and pocketbooks of the rich was the poster. This was especially true in Paris, where bills were used to advertise everything from art shows to automobiles. Posters often popularized talented but unknown artists, who could eat for three months on the proceeds from one good design. Among these unknowns was Alfons Mucha, who began his career with commissions from Sarah Bernhardt *(right)*. Posters became such a rage that a black market for them developed, and some art dealers even advised their clients to sell a Rembrandt to buy a Mucha or a Toulouse-Lautrec.

SARAH BERNHARDT *posed for Alfons Mucha to advertise a pla*

A BEMUSED DRINKER *stares out from a poster by Jacques Villon.*

A PRIZE-WINNING RACING CAR, *led by a wind-blown spirit of speed, was featured in this Parisian advertisement by Eugène Verneau.*

A LANGUID LADY *admires "Job" cigarette papers in another Mucha poster.*

AN ART SHOW *inspired this poster from Vienna, where Art Nouveau took on modern overtones.*

A BREAKING WAVE *was seen as a row of leaping horses by Walter Crane, who utilized Art Nouveau's curving lines.*

A LINEAR WOODCUT, *one medium which appealed to Art Nouveau designers, was made by the Norwegian artist Edvard Munch. Its flat surface and free lines are characteristic of the style, but the specter-like face reflects Munch's personal interpretation of his subject.*

AN ETHEREAL PORTRAIT *of a Viennese lady by Gustav Klimt displays strong influences of the new style. The glittering, ornamental surface, the preoccupation with detail, and the dreamy quality of the figure are all more suggestive decoration than of a real person*

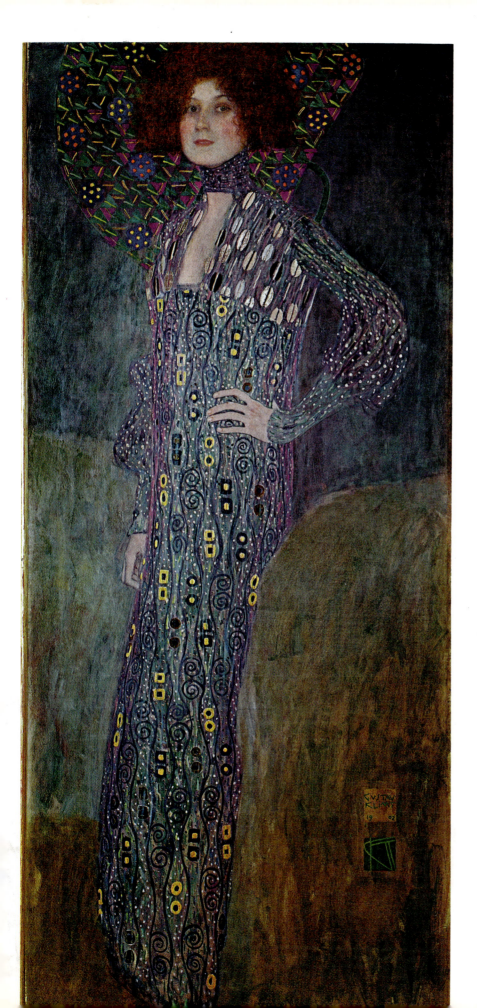

SUBTLE EFFECTS ON PAINTING

Art Nouveau flourished in the graphic and decorative arts, but never became an identifiable style of painting. It exerted only subtle influences on the canvases of the day, because at the turn of the century painters were too busy experimenting with a variety of other new styles such as Expressionism and Cubism. A few characteristics of Art Nouveau, however, gradually began to appear even in the work of such masters as Seurat and Picasso.

Probably the painter who adhered most rigidly to the principles of the new style was Gustav Klimt, an Austrian artist who painted the portrait shown at the left; in fact, he concentrated so much on surface ornamentation that the woman's face became secondary. In the manner of some other Art Nouveau enthusiasts, Klimt was obsessed as much with his personal appearance as he was with that of his work. He often appeared at fashionable gatherings bearded and robed like a Messiah, and adorned with epaulets embroidered by society ladies who were his devotees.

THE ULTIMATE
FLOURISH

If any one art can be said to have embodied all the ideals of Art Nouveau it was architecture, particularly the work of the Spaniard Antonio Gaudí. Gaudí's imagination was completely uninhibited; even the roof of his Barcelona apartment house —built in 1910 and still standing—swooped and swirled in the

best Art Nouveau manner *(above)*. Chimneys loomed as weird, monster-like forms, and turrets and towers encrusted with mosaics sprouted everywhere. Every window was different, each one virtually a piece of sculpture.

Gaudí's work, like the rest of Art Nouveau, was immensely popular in its time. But in the early decades of the 20th Century, the style suddenly waned. Launched as a reaction to the stilted classicism of the 1900s, it was rejected in turn by a new generation of artists who objected to its very opulence. In its place came an austere new movement called Modern Art.

8

THE CRUSHING FINALE

The passage of the 19th Century into the 20th was marked by every evidence of harmony and peace—and above all by hope. A whole generation had come of age without witnessing a major conflict of arms, and many supposed that war, like the duel and the feud, was a barbaric convention that civilized man had outgrown. All visible signs seemed to indicate that progress and prosperity were limitless, that poverty and disease would cease to be serious problems, and that man would evolve comfortably toward a higher civilization.

Today that turn-of-century era appears as an Indian summer, a period of warmth and mildness that marked the close of an epoch. There was, in fact, a good deal of tension concealed beneath the surface calm of European civilization. Ferment in social, political and intellectual life had already caused outbursts whose significance is now clear. But most people of the period complacently assumed that the challenges to the established order would be met in good time.

The fact is that almost every beneficent aspect of progress had a formidable side. Nationalism had melded peoples together in political unity; at the same time it had fanned discontent among those whose aspirations had not yet been achieved. The acquisition of colonial territories had opened the markets of the world to Europe; at the same time it had embittered competing nations, and the resulting antagonism helped to forge powerful alliances among them. These combustible situations were to lead to a disastrous explosion. The era that people called an age of progress would soon end with one of history's bloodiest wars, in which millions of young men would lose their lives and in which several governments would fall.

The signs of trouble, which first became evident sometime in the 1890s, went largely unnoticed for several years. A traveler through Europe as the 19th Century neared its end would have found the same mood prevalent in nation after nation.

In 1897 Great Britain enthusiastically celebrated 60 years of the reign of Queen Victoria. In London that year there was a Thanksgiving Service at St. Paul's Cathedral. The festivities opened with a stately procession led by a certain Captain Ames

FASHIONABLE DINERS *in evening clothes reflect the opulence of Europe on the eve of the horrors of World War I. In this painting, elegant Parisians gather for dinner at the Pré-Catelan restaurant in the Bois de Boulogne.*

who was six feet eight and the tallest man in the British Army. Behind him came a file of carriages of state, then mounted cavalry, then men on foot in the varied uniforms of Britain's far-flung empire: Canada, India, Australia, Trinidad, Borneo, Jamaica, Nigeria, Hong Kong, Singapore, Sierra Leone. At the end of the procession, in an open carriage drawn by eight cream-colored horses, rode a small figure in black: the 78-year-old Victoria, Queen of the United Kingdom of Great Britain and Ireland and Empress of India. Along six miles of London streets bedecked with flowers, millions of onlookers cheered in an outpouring of national pride.

This colorful pageant was more than a tribute to the person of the Queen; it was a proud display of the British nation, a celebration of its extensive empire and its bounteous wealth. In the Jubilee Summer of 1897 Great Britain was the richest of the great powers ruling the world, and she was happily showing off her affluence.

Victoria reigned for four years more. After her death in 1901 her eldest son became King Edward VII at the age of 59. Edward was shrewd, patriotic and European-minded; he was also a polished man of the world who enjoyed pomp and luxury, gaiety and self-indulgence. His reign lasted less than a decade, but this jovial monarch gave his subjects an added sense of the good life and the inevitability of progress. Englishmen emulated their sovereign, and enjoyed their riches with careless abandon.

France was equally gay. By the last decade of the 19th Century the terrors of the Franco-Prussian War and the Paris Commune seemed far behind. The Third Republic, created in 1870, had proved itself politically stable and economically sound; in one generation it had raised France to a position of world influence. Overseas the French had acquired a bountiful empire of some three million square miles in Africa and Southeast Asia.

Paris shone as one of the great cultural centers of Europe. Connoisseurs of art no longer thought it amusing to ridicule the Impressionist painters; Manet had received the cross of the Legion of Honor, and others were respected among painters the world over. Scholars looked to the Sorbonne as one of the great institutions of European learning. A French neurologist, Jean Martin Charcot, was investigating the human nervous system and hysteria; his findings contributed to the development of psychoanalysis by his pupil Sigmund Freud. In 1896 the physicist Antoine Henri Becquerel discovered radioactivity in uranium; his protégés Pierre and Marie Curie later isolated radium from pitchblende.

Nowhere was the sense of accomplishment stronger as the 19th Century drew to a close than in Germany, where the Kaiser proudly asserted in 1896 that "the German Reich has become a *Weltreich*," a world power. The unified nation's commerce and industry had come almost abreast of Britain's. German bankers had interests in Latin America and in the Orient. German engineers were in the process of building the so-called Berlin-Baghdad railway, intended to link Europe with the Persian Gulf—a prospect that England, Russia and France viewed with dismay, for it would draw the Middle East into the German sphere of influence. Though the Germans had been late to enter the imperialistic race for colonial territories, by 1900 they had some holdings in Africa and the Pacific. They also had the strongest standing army in Europe.

The German throne was occupied by Wilhelm II, a swaggering monarch who conceived of himself as the deputy of Providence. He had undergone a rigorous military training, and despite the handicap of a withered arm he learned to swim and ride horseback, even to hunt and play the piano. Though he was autocratic in the extreme, his subjects found him thoroughly attractive. But there was one disquieting sign for those with the vision to detect it. The writings of the philosopher Friedrich Nietzsche, then fashionable in intellectual circles, envisioned the evolution, through the discipline of body and intellect, of a Superman; the idea was being perverted in Germany into a glorification of brute strength and national superiority. By 1900 the Germans, claiming the world's mightiest military machine, were aggressively preparing to secure for themselves their "rightful place in the sun."

Behind these countries in wealth, but equally hopeful of progress, was the Austrian Empire, where the Habsburg monarch reigned over a collection of subordinate peoples in whom nationalist sentiment grew fiercer every year. The Emperor, Franz Joseph, who was to hold the throne for a total of 68 years, until he reached the age of 86, was a man of spartan discipline and feudal disposition. From 4:30 a.m., when he arose, until late at night he worked at his desk, taking his meals from a tray and allowing himself few diversions aside from an occasional hunt. He was autocratic and resistant to the encroachment of democracy, but he was not unable to compromise, nor was he blind to modern trends. "You see in me the last monarch of the old school," he said one day to Theodore Roosevelt. He later remarked, "I have known for a long time how much of an anomaly we are in today's world."

His Empire was as anomalous as he. By the end of the 19th Century Austrian industry and capitalism existed side by side with traditional crafts and primitive agriculture. Although the nation was run by a powerful and conservative middle class, it could boast of considerable social legislation. Austrian workers had acquired the right to collective bargaining as early as 1870; by 1911 some four million were covered by workmen's compensation.

Just as certain concessions had been made to social progress, so there had been some recognition of nationalist claims. In 1867, after the loss of his Italian territories, and in an effort to appease the national aspirations of the 10 million Hungarian Magyars under his sovereignty, the Emperor had divided his realm in two. Franz Joseph remained as ruler of both halves, taking the titles of Emperor of Austria and King of Hungary, and presiding over common ministries of military, foreign and fiscal affairs. But the Empire acquired two constitutions, two parliaments, and two capitals, one in Vienna, the other in Budapest.

The split was essentially an agreement between the Germans and the Magyars for joint domination of the other peoples who lived in Austria-Hungary; these included six million Czechs and 20 million Poles, Ruthenians, Romanians, Croats, Slovaks, Slovenes and Serbs. "You look after your barbarians," the Austrian Chancellor Beust is said to have told the Hungarian official attending the settlement, "and we will look after ours." Together the minorities outnumbered their German and Magyar rulers. They were collectively Slavs, and therefore kin to the Russians—a fact that was to have a crucial significance in world affairs.

Both halves of the Empire had parliamentary machinery, but that machinery made a mockery of representative government. The Germans constituted only 35.6 per cent of the Austrian population, but they controlled the Parliament because the electoral system favored the wealthy, who were predominantly German. Even those who sat in the Parliament had an uncertain voice in affairs of state; the Emperor and his ministers often evaded their legislation. In Hungary the situation was even worse. The franchise had educational and financial qualifications that restricted the ballot to 6 per cent of the population, and this 6 per cent consisted entirely of the Magyar gentry. Some 95 per cent of all state officials, 92 per cent of all county officials and 90 per cent of the judiciary were Magyar.

The compromise was an uneasy one, especially as the Slavic minorities began to acquire a higher standard of living. They grew increasingly insistent on recognition, and their agitation—encouraged by the neighboring Serbs, who were also Slavic—was to rend the Empire asunder.

Vienna, the primary capital of the Empire, was a city of great beauty, of cultural attainment and of gaiety. Spread out on the south bank of the Danube river where it enters the Hungarian plain, lying between tall hills and surrounded by beech trees and meadows, it was unsurpassed in natural setting. The Ringstrasse, which girdled the center of town, was made by 19th Century city planners into one of the loveliest streets in Europe—two and a half miles in circumference, 200 feet wide, and planted with rows of lindens and plane trees. Vienna was also the cultural and financial Mecca of the Empire, a center for doctors and lawyers, painters, musicians and writers, and businessmen.

If the Emperor was spartan, his Viennese subjects were epicurean. Vienna was a city of music, laughter and flippancy. A 19th Century writer recorded: "The people of Vienna seem to any serious observer to be reveling in an everlasting state of intoxication. Eat, drink and be merry are the three cardinal virtues and pleasures of the Viennese. It is always Sunday, always carnival time for them."

Although the Viennese were outwardly gay, they were inwardly nervous. It is noteworthy that Sigmund Freud was a product of the Vienna of this period. It was here that he probed the unconscious recesses of human nature, finding uncertainty and uneasiness underneath the surface of frivolity.

The Viennese were not the only people growing nervous as the century turned. Throughout Europe the strains were increasing every day, domestically and internationally, among statesmen and among the people. Nationalism was growing heated, causing divisiveness and contention where once it had been a force for cohesion. Common enmities were driving compatible nations into alliances, and their adversaries into matching alliances; the resulting hostility and fear led to arming in preparation for an undefined war that statesmen began to think might come. Serious trouble was brewing on the borders of industrial Europe, in the Balkan peninsula, and although it first seemed remote it was to be disastrous: it was to draw the nations of Europe into worldwide conflagration.

The forging of the modern alliance had been begun in Germany by Bismarck, who in 1879 formed an agreement with Austria-Hungary by which each government promised to support the other in case of attack by a third power. This became the Triple Alliance in 1882, when Italy was admitted to the pact. Against this combination of power, France signed a similar agreement with Russia in 1894; by 1907 Britain, traditionally isolationist, had joined them both in the Triple Entente. The six great nations of Europe now faced each other in two massive blocs. These were loudly proclaimed to be defensive rather than offensive measures; nevertheless both camps armed themselves heavily.

Never in peacetime had nations maintained such vast armies as at the beginning of the 20th Century. Germany, which had found an army a useful instrument of power during its struggle for unification, kept a standing body of troops in existence even after 1871 because Bismarck feared his newly created state might otherwise disintegrate. Germany enforced universal conscription, requiring every able-bodied youth to serve for two years, and she increased the size of her army in synchronization with the strains on European international relations. Her forces were raised from 400,000 in 1871 to 557,000 in 1894; to 605,000 in 1904; to 820,000 in 1914. France matched every raise, increasing her army in that period to 750,000 men. Austria and Italy followed suit. The only apparent exception was Britain—but Britain had the Royal Navy.

Suddenly Britain's naval supremacy confronted its stiffest challenge in centuries. Germany, pos-

This accumulation of armies and arms, which was inspired by fear of enemies, was accompanied by the spirit of nationalism, a sentiment that affected the masses and statesmen alike—and all nations, the great as well as the small.

Nationalist sentiment gave the lonely individual in an urbanized society a sense of belonging—and often it gave him a mistaken sense of national superiority as well. All the manifestations of progress—governmental and social reform, industrial growth, literary and artistic flowering and imperial acquisition—were hailed as national achievements and touted as distinctions that set one's own nation apart from all others. As literacy increased, this sentiment was fed by the yellow press; as leisure increased, it was even fed by the stage. An English music hall singer, Gilbert Hastings Macdermott, gave a word to the language and fuel to the fire of nationalism when he sang:

We don't want to fight, but, by jingo, if we do,
We've got the ships, we've got the men, and got
* the money too!*

From this verse came the word "jingoism," meaning bellicose patriotism. And as the lyric implied, the means of aggression were available to all.

As the two alliances grew stronger and more suspicious of each other, a series of diplomatic crises occurred, each more difficult to resolve than the last, and each adding its bit to the mounting strain. In 1905, thinking to jar the Anglo-French Entente Cordiale, the Kaiser delivered an oration in Morocco, where the French had already established a foothold, urging independence for that territory and demanding an international conference to discuss the issue. Fearing Germany's massive strength and diplomatic blustering, the major powers agreed to convene—but to Germany's surprise the conference succeeded only in tightening the bond between Eng-

sessing a short coastline and few overseas colonies, had never had a navy to speak of. But in 1897 the Kaiser decided that Germany must have a mighty fleet to rival Britain's. He accordingly laid plans for an ambitious naval building program. For the British, an island people dependent on imports, naval security was a matter of life and death. They countered the German naval buildup in 1906 by launching H.M.S. *Dreadnought*, the greatest battleship ever constructed up to that time. She had 11-inch armor plate, ten 12-inch guns, and turbine engines that could push her through the seas at an unprecedented 21 knots. She was twice as formidable as any ship then afloat. By 1914 Britain had 19 battleships of the *Dreadnought* class at sea and 13 building. Germany, struggling to catch up, had 13 similar vessels afloat and seven in the yards.

land and France; only Austria sided with Germany, and France retained her prerogatives in Morocco. A second crisis developed over Morocco in 1911, when a German gunboat arrived at the port of Agadir "to protect German interests." This was international blackmail, for Germany had no significant interests in Morocco, whereas France already had plans for acquiring a protectorate there. Once again a nervous France submitted to the threat: Germany promised no further harassment in exchange for a strip of territory in the French Congo. Once again the French and English ties were reinforced.

In the Balkan peninsula, by the early years of the 20th Century, crises were occurring so regularly as to be a part of the region's way of life. The Balkan peninsula was a congeries of territories, some of them independent kingdoms, some of them under Turkish suzerainty for nearly 600 years. They were economically backward, politically unstable, and out of the mainstream of European industrialization. But their affairs were of moment to all of Europe. Passage through them was necessary to those nations that traveled to the Mediterranean, either overland or by way of the Black Sea, to trade with the East. The politics of the Balkan peoples were of particular concern to Russia, which sympathized with their drive for Slavic nationalism, and to Austria, which strove to thwart that drive in her own reach for power. Both nations consequently interfered in Balkan affairs, now helping, now frustrating, the activities of this or that faction. Russian and Austrian meddling, in turn, was eyed with suspicion by France and England on the one hand, and Germany on the other, with a view to its effect on the balance of power.

The first crisis to arouse international indignation over the Balkans occurred early in 1908, when Austria appropriated from Turkey the territories of Bosnia and Herzegovina, which lay south of the Empire and adjoined the independent kingdom of Serbia. Serbia was outraged; she was the major center of Slavic nationalism south of Russia, and had hoped to make Bosnia and Herzegovina part of a great national Slavic state that she hoped some day to form. Russia was equally angry. England and France as well were alarmed at the Austrian show of strength—but when Germany insisted that Russia acquiesce to the Austrian move or face war, all three members of the Entente subsided.

There was a reason why Austria was able to accomplish the coup in Bosnia and Herzegovina so easily: the 600-year-old Turkish Empire was nearing collapse—and the Austrians were not the only ones to notice. The independent Balkan nations, some of which had suffered under Turkish oppression for hundreds of years, now girded themselves to prey on Turkey's weakness. The result was a series of confused incidents in which enemies formed brief liaisons, allies fell out and sparred with each other—and the world was propelled relentlessly toward a major conflict.

In October 1912 Serbia and Montenegro (both of which are part of modern Yugoslavia) joined with Bulgaria and Greece to declare war on Turkey. In two months these small countries surprised the whole world by driving the Turks out of all of Europe except Constantinople and the Dardanelles. They reached an armistice with Turkey in December and eagerly prepared to divide the spoils.

Here the great powers, sensing trouble, stepped in. Austria could not tolerate the rise of strengthened Slavic states on her borders. Russia, on the other hand, could not see Slavic gains repressed. England and France could not sit still if Austria, with Germany behind her, moved into the Balkans; this would alter the balance of power. Even Italy spoke up; she fancied the Adriatic as an Italian lake, and Serbia was planted on its eastern shore. In an effort to preclude a major international conflict, all these powers decided to try resolving their differences around a conference table. They summoned the Balkan victors to a meeting in London before they could act on their own.

In London the great nations imposed a series of arrangements almost comically inappropriate. Serbia failed to get the section of Adriatic coastline she had won; instead she got a piece of inland territory that Bulgaria wanted. Out of the Adriatic strip was created a new state, Albania, which included territory that Greece had coveted. In compensation Greece got some lands along the Aegean coast—and this disgruntled Russia.

Not surprisingly, the London settlement satisfied no one. One month later the erstwhile Balkan allies had turned upon each other and started a local war. This time, however, they resolved their own quarrel by themselves, at the Treaty of Bucharest in August 1913. The boundaries on which they decided at Bucharest have, with one exception, endured to this day. The borders of the major nations, in contrast, have altered considerably more over the same period of time.

Serbia emerged from the wars with the greatest gains. Though she had failed to acquire the Adriatic coastline she so much wanted, she had nearly doubled her territory and increased her population by 50 per cent. Now that the Turks were virtually expelled from Europe, Austria remained the chief obstacle to fulfillment of Serbian aspirations, and this hurdle Serbia meant to overcome. As the conference at Bucharest was breaking up the Serbian Prime Minister was heard to say: "The first round is won; now we must prepare the second against Austria." A few days later, having a rest at Marienbad, he remarked to a fellow Serbian: "Already in the first Balkan War I could have let it come to a European war, in order to acquire Bosnia and Herzegovina."

It was now clear that European peace was in real danger. An American visiting Europe some months later, in the spring of 1914, wrote: "The whole of Germany is charged with electricity. . . . It only needs a spark to set the whole thing off." The German Ambassador in Paris said, "Peace remains at the mercy of an accident."

The accident came shortly thereafter, in Sarajevo, a small town in Bosnia to which the Archduke Franz Ferdinand, nephew of Franz Joseph and heir to the Habsburg crown, went on an ill-advised inspection tour. When the Archduke's visit to Bosnia was announced, Serbs all over the world sounded a call for revenge.

"Serbs, seize everything you can lay hands on," exclaimed a Serbian newspaper in Chicago—"knives, rifles, bombs and dynamite. Take holy vengeance! Death to the Habsburg dynasty, eternal remembrance to the heroes who raise their hands against it!"

In Belgrade, the capital of Serbia, several secret terrorist societies were already at work training youths in the use of arms. When the Archduke's visit was imminent, one of these societies, the Black Hand, sent three 19-year-old boys across the border into Bosnia armed with guns and bombs and bent on assassination.

On the morning of Sunday, June 28, the Archduke was to drive down Franz Joseph Street in an open car. Before he arrived a hostile crowd had gathered and one bomb had exploded. Military officials, concerned, gave last-minute orders to alter the royal route, but by the time word reached the Archduke's chauffeur, he had already entered Franz Joseph Street. He stopped the car. The side-streets were too narrow for passage, so he had to back up. But of all the spots to pick in a route that was several blocks long, he had halted directly in front of one of the three assassins of the Black Hand. While he shifted gears one of them fired two shots, wounding the Archduke and his duchess. As blood spurted from his mouth, Franz Ferdinand turned to his wife. "Sopherl, Sopherl!"

he cried. "Don't die! Stay alive for the sake of the children." Then he collapsed on the seat, saying over and over again, "It is nothing. It is nothing." He was dead a few minutes later.

It was this fanatic act by a group of Serbian nationalists in an obscure and insignificant town that proved to be the crisis European diplomacy could not meet. All nations were outraged by the deed, but for weeks none took action. The government of Austria-Hungary waited, largely because Franz Joseph wanted assurance of German support for whatever he did. Then, on July 23, 1914, after four weeks, his foreign minister, Count Leopold von Berchtold, delivered an ultimatum to Serbia, demanding satisfaction within 48 hours. One of the stipulations was that all nationalist agitation in Serbia must cease; another was that Austrian officials must see to its enforcement.

Serbia replied in the required 48 hours, in a conciliatory tone and accepting all the conditions except the demand for Austrian supervision; this amounted to a surrender of Serbian sovereignty, and this Serbia could not countenance. On hearing the extent to which the Serbs had capitulated to the Austrian demands, the Kaiser sent congratulations on Austria's "brilliant performance" and "great moral victory." But Berchtold, who had obtained a promise of military aid from Germany, pushed for full acceptance; when Serbia stood fast Austria declared war. The date was July 28.

The Russians, still smarting over the humiliations of 1908 and 1912-1913, and determined to strike a blow for Slavic aspirations in Europe, immediately mobilized to go to the aid of the Serbs. France, Russia's ally, followed suit. On August 1 Germany declared war on Russia and two days later on France. A day after that, England, the third member of the Triple Entente, declared war on Germany. The First World War had begun.

Few foresaw the consequences. The English thought the fighting was merely a spat that would be over by Christmas, and the German Crown Prince looked forward to a "bright and jolly war." Instead it grew into a four-year holocaust that engaged 65 million men, and cost the lives of almost 10 million of them. It was to bring about the collapse of several governments and the demise of three dynasties—the Hohenzollern in Germany, the Romanov in Russia and the Habsburg in Austria —the youngest of which dated back 300 years. It was to impoverish most of Europe for years to come. Thus the Age of Progress reached a disastrous finale.

The eruption of war was the culmination of a series of shocks dealt to society in the aftermath of the Industrial Revolution. Earlier ages had seen calm and steady change—change as slow and imperceptible as geological erosion and biological evolution. With the Age of Progress change became dynamic and ever-accelerating, like the machinery that was its hallmark.

Yet, despite its closing tragedy, the era was truly an age of progress. Ideas that 20th Century Western civilization takes for granted—the value of science and technology, the virtues of representative government, the necessity for education of the masses—all these antedate the war; most of them came to the forefront during the latter half of the 19th Century. Thanks partly to the industrial miracle, partly to gains in education, man had acquired the power to manipulate his own environment. Of all the gains made during the era, perhaps the most striking were in education. It was during the Age of Progress that nations began sending whole populations to school, and hence that for the first time in Western history the means for self-improvement—social, economic, political and cultural—became available not solely to the well-born and the fortunate, but to everybody. This is perhaps the most precious legacy of the Age of Progress.

CAMERA BUFF EMILE ZOLA, *who was fascinated by photography, posed while taking a picture himself.*

ALBUM OF AN ERA

The Age of Progress was a vigorous and flamboyant epoch, and its heroes and heroines were an appropriately colorful gallery of individualists: writers, artists, actors, monarchs, scientists, inventors, entrepreneurs, geniuses, libertines, rogues. Such an array of personalities was not to be seen again in a world increasingly occupied with the sober realities of World War I and its aftermath. While they held the spotlight, most of these figures were captured by the camera, a new rage of the day. Many of them, indeed, were avid photographers themselves—like novelist Emile Zola (*above*), who snapped more than 6,000 shots.

Tschaikovsky at 23, as a Ministry of Justice clerk

Alexandre Dumas with actress Adah Menken

Nijinsky in "The Faun"; it shocked even the French

Oscar Wilde, man of letters, in a languid mood

Pavlova, solo ballerina of "The Dying Swan," with her pet

Victor Hugo, who once captioned a photo: "Hugo listening to God"

Henrik Ibsen, "father" of modern drama

Playwright Anton Chekhov (left) and novelist Maxim Gorky; they dramatized the many ills of Russian society in the days of the czars

GREAT NAMES IN MUSIC, LITERATURE AND DRAMA thrilled the public: the composers Peter Ilyich Tschaikovsky and Franz Liszt; Victor Hugo, author of *Les Misérables*; the ballet immortals Vaslav Nijinsky and Anna Pavlova. The lusty off-stage antics of some of the era's figures brought them a notoriety that rivaled their artistic fame: Oscar Wilde was jailed on a morals charge and subsequently wrote his *Ballad of Reading Gaol*. Alexandre Dumas, author of *The Three Musketeers* and some 200 other historical romances, boasted that he had fathered 500 children. The American actress Adah Isaacs Menken, another freethinker, announced publicly that she was going to Paris to become Dumas' mistress—and did.

Virtuoso Franz Liszt: when he played the piano, women shrieked

SCIENTISTS AND TECHNICIANS joined hands to revolutionize life in the late 19th and early 20th Centuries. Almost as fast as the former made discoveries, the latter put them to use, frequently making fortunes in the process. Many advances were achieved by men with little or no technological training: Ferdinand de Lesseps, a French promoter, built the Suez Canal; Louis Daguerre, a painter, concocted the fixative that kept photographs from fading; John Boyd Dunlop, a Scottish veterinarian, invented a pneumatic tire so his son's tricycle wouldn't give him such a bumpy ride.

Louis Blériot (left, in helmet): he flew across the Channel to England in 1909

Camille Flammarion (below), popularizer of astronomy

Tower-builder Eiffel and children

Melli Beese, Germany's first licensed aviatrix

Canal-builder Ferdinand de Lesseps, out for a drive with seven of his 17 children

John Dunlop, taking a spin on his tires

Engineer I. K. Brunel, with the giant chains used to launch his "Great Eastern," five times bigger than any ship afloat

Marie Curie, co-discoverer (with her husband Pierre) of radium

ologist Thomas Huxley, who converted England to Darwin's theory of evolution

Jean Corot, who quit his studio to paint landscapes outdoors

Auguste Rodin, the most famous sculptor of his day

Henri Rousseau, a customs inspector—and a brilliant Sunday painter

Edouard Manet, one of the early practitioners of Impressionism

Toulouse-Lautrec, in Japanese costume for a party

Gustav Klimt, Viennese leader of Art Nouveau

ADVENTUROUS ARTISTS led 19th Century painting and sculpture out from behind a stylized, romantic veil and began interpreting the world as they saw it. Gustave Courbet, founder of the "Realist" school, turned to such everyday scenes as peasants at a funeral and workmen breaking rocks. Taking Realism into a new dimension, men like Edouard Manet, Auguste Renoir and Paul Cézanne caught brilliant, momentary impressions of contemporary life with dramatic new techniques of color and light. Out of the bold leadership of such men grew many schools of art, and through their efforts and insights they gave the age a remarkable picture—a glimpse into its own soul.

Realist painter Gustave Courbet

Paul Cézanne, perhaps the age's greatest artist, with one of his famous "bathers" scenes

Paul Gauguin, trouserless, at the harmonium

Impressionist Auguste Renoir and wife

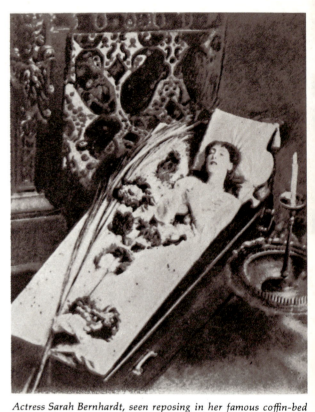

George Sand: she wrote 80 widely popular novels

An array of exciting women gave the era much of its color and verve—and the public plenty to talk about. Sarah Bernhardt, the most celebrated actress of the epoch, startled her admirers by sleeping in a coffin. George Sand, nee Amandine Aurore Lucie Dupin, was as well known for her affairs as for the novels she wrote. The lovely Evangeline Booth was more virtuous in her fame: she remained a Salvation Army lassie, refusing the hand of a rich Russian prince.

Actress Sarah Bernhardt, seen reposing in her famous coffin-bed

The Five Barrisons, dressed in little-girl clothes for their "kittenish" vaudeville act

Evangeline Booth, daughter of the Salvation Army's founder, with two waifs

Paris dancer Cléo de Mérode

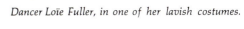

Dancer Loïe Fuller, in one of her lavish costumes.

Gaby Deslys, friend of kings, and her limousine

Opera singer Lina Cavalieri

oprano Mary Garden, in yard-long tresses as Mélisande

Mata Hari: she failed as a dancer, then as a spy, was shot in 1917

THE CROWNED HEADS OF EUROPE never glittered more brilliantly than they did in the Indian summer of kings, just before World War I. Queen Victoria's dream of giving birth to a closely knit royal family that would occupy the palaces of Europe seemed to be coming true: she had produced 15 children and grandchildren who either ruled, or married the men who ruled, Russia, Germany, Greece, Romania, Spain, Norway and smaller states. But already the death knell was tolling for royalty. Among those who heard it was Czar Nicholas of Russia. After the assassination of Austria's Archduke in 1914 he wired his cousin Kaiser Wilhelm of Germany: "Very soon I shall be overwhelmed by the pressure brought upon me." Four years later the Czar and his family were killed in a Siberian cellar during the Russian Revolution. The curtain had fallen on a king, as it was falling on a spectacular age.

Victoria, with a picture of her beloved Albert

Wilhelm II (left), parading with his six sons up Unter den Linden in Berlin in 1913

Britain's Edward VII, in his honorary German general's uniform

Austria's Archduke Rudolf in hunting dress

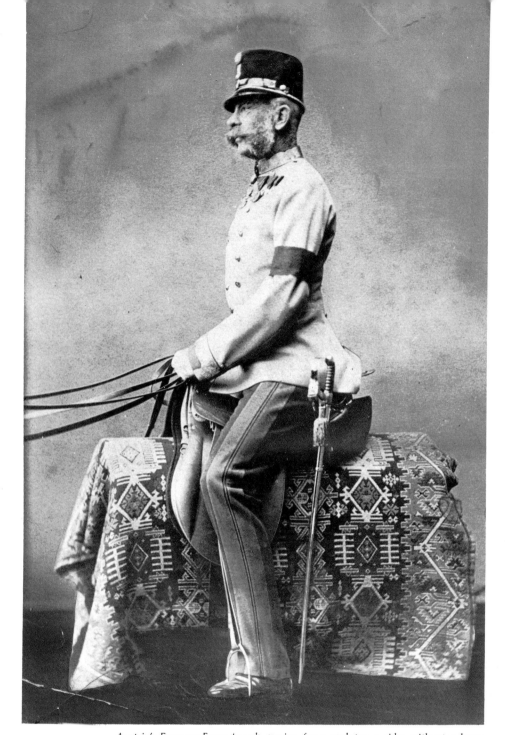

Austria's Emperor Franz Joseph, posing for a sculptor—a rider without a horse

pain's Alfonso XIII, with his young son

Russia's Czar Nicholas II, boating with his son Alexis on a holiday

CHRONOLOGY: *A listing of events significant in the history of the Age of Progress*

Politics and War

Science and Technology

1850

1851	Louis Napoleon seizes power in France
1852	France's Second Empire begins
1854	The two-year Crimean War starts
1858	Count Cavour plots with Napoleon III to drive the Austrians out of Italy

1851	The Great Exhibition opens in London's Crystal Palace
1856	Henry Bessemer invents his converter to make pig iron into steel
1857	Louis Pasteur begins his pioneering study of fermentation
1858	Rudolph Virchow publishes his work on cellular pathology
1859	Charles Darwin completes the *Origin of Species*

1860

1860	The Kingdom of Italy is established
1861	Napoleon III begins his six-year campaign to conquer Mexico
1862	Bismarck becomes Minister-President of Prussia
1864	Prussia declares war on Denmark over Schleswig-Holstein
1866	The Seven Weeks' War is fought at the Battle of Königgrätz
1869	The Social Democratic Working Men's Party is founded in Germany

1861	Ernest Solvay patents a soda-making process which drastically reduces the cost of manufacturing textiles, glass and soap
1864	Siemens and Martin introduce the open-hearth process for making steel
1865	Joseph Lister introduces antiseptic surgery
1866	Dynamite is patented by Alfred Nobel
1866	The first transatlantic cable is laid
1867	Werner Siemens introduces his dynamo for generating electricity

1870

1870	The Franco-Prussian War begins
1871	The Paris Commune seeks to set up a revolutionary government in Paris
1871	The German Empire is established; France forms its Third Republic
1873	The Three Emperors' League is formed by Austria, Russia and Germany
1876	Alfonso XII ascends the Spanish throne
1877	The Russo-Turkish War begins
1878	The Congress of Berlin partitions Africa among the European powers
1879	The Dual Alliance is formed between Germany and Austria-Hungary

1871	Dmitri Mendeleev discovers gallium to add to his Periodic Table of the Elements
1871	Charles Darwin publishes his controversial *Descent of Man*
1873	Clerk Maxwell's study on electricity and magnetism appears
1876	Nikolaus Otto develops the four-stroke internal combustion engine
1878	Pasteur lectures on his germ theory at the Academy of Medicine in Paris

1880

1881	The Emperor Alexander II is assassinated by "The People's Will," a Russian terrorist society, and Alexander III ascends the throne
1882	Italy joins Germany and Austria-Hungary to form the Triple Alliance
1883	Bismarck establishes social welfare measures in Germany
1888	German Emperor Wilhelm I dies and is succeeded by Wilhelm II

1882	Robert Koch isolates the bacillus of tuberculosis
1884	Charles Parsons develops a steam turbine
1885	Hiram Maxim invents the machine gun
1886	Aluminum is made by an electrolytic process for the first time
1888	John Dunlop produces his pneumatic tire

1890

1890	Bismarck is dismissed, assuring Hohenzollern authority
1890	Felix Méline, French cabinet head, puts through a customs tariff in France
1893	An independent labor party is established in England
1894	The Dreyfus case begins
1899	The first Hague Peace Conference is held

1890	An all-steel bridge is completed over the Firth of Forth
1897	Rudolf Diesel patents the diesel engine
1898	Radium is discovered by the Curies

1900

1900	Italy's King Humbert is assassinated and is succeeded by Victor Emmanuel
1901	Queen Victoria dies and King Edward VII takes the throne
1903	The Russian socialist groups, Bolshevik and Menshevik, split over doctrine
1904	The Anglo-French Entente is created
1904	The Russo-Japanese War begins
1905	Revolution breaks out in Russia
1907	The Triple Entente is created
1908	The Young Turks' Revolt seeks to rejuvenate the Ottoman Empire

1900	Max Planck presents his quantum theory
1901	The first message is sent over Marconi's transatlantic wireless telegraph
1903	The Wright brothers make their first airplane flight
1905	Einstein presents his theory of relativity
1907	Bergson publishes *Creative Evolution*
1909	Blériot makes the first airplane flight across the English Channel

1910

1910	Portugal's monarchy is overthrown
1911	The Agadir Crisis takes place in Morocco
1912	The Balkan wars begin
1914	The Archduke Franz Ferdinand is assassinated at Sarajevo
1914	The First World War begins

1910	Russell and Whitehead write *Principia Mathematica*
1911	Ernest Rutherford creates a nuclear model of the atom

Expansion and Colonization

Art, Literature and Music

1850

1853 Commodore Perry opens Japan to East-West trade

1857 The Anglo-French War against China is ended by the Treaty of Tientsin, which confers economic rights on foreigners

1857 Gustave Flaubert publishes *Madame Bovary*
1857 John Stuart Mill publishes his essay "On Liberty"
1857 Free libraries are opened in England and Germany

1860

1860 Russia founds the Far Eastern city of Vladivostok

1860 Victor Hugo writes *Les Misérables*

1862 The first French annexations in Cochin China take place
1864 A British trading company completes the first Persian telegraph

1863 Edouard Manet paints his famous *Déjeuner sur l'herbe*

1866 Dostoevsky publishes *Crime and Punishment*

1869 The Suez Canal is opened, cutting the distance between Europe and the East

1867 Karl Marx completes Volume I of *Das Kapital*

1870

1871 Stanley finds Livingstone in Africa

1871 Emile Zola begins *Les Rougon-Macquart*
1871 Giuseppe Verdi completes the opera *Aïda*
1873 Herbert Spencer completes his *Study of Sociology*

1875 Great Britain purchases controlling shares in the Suez Canal

1874 Wilhelm Wundt publishes *Foundations of Physiological Psychology*

1877 Rodin exhibits his sculpture for the first time in Paris

1878 The International Congo Association is formed by King Leopold of Belgium, H. M. Stanley and private financiers

1879 Ibsen writes *A Doll's House*

1880

1881 The French occupy Tunis

1880 Dostoevsky publishes *The Brothers Karamazov*

1882 The British occupy Egypt

1882 Wagner produces *Parsifal*
1883 Nietzsche publishes *Thus Spake Zarathustra*

1884 Germany establishes colonies in East Africa; trade unions are legalized in France
1885 Bismarck calls the Berlin Conference on Africa
1886 Burma is annexed by the British

1886 Kipling publishes *Departmental Ditties;* Nietzsche's *Beyond Good and Evil* appears
1887 Strindberg writes *The Father*

1890

1893 Tschaikovsky writes his *Symphonie Pathétique*

1896 The Italians are defeated in Ethiopia

1896 Lenin publishes *Development of Capitalism in Russia*
1897 Cézanne paints *Lake of Annecy*

1899 The Boer War starts in South Africa between English and Afrikaners
1899 The Boxer Rebellion breaks out in China, and is put down by the European powers, Japan and the United States

1899 Freud makes public his controversial *Interpretation of Dreams*

1900

1901 Thomas Mann writes *Buddenbrooks*

1904 The Trans-Siberian Railroad is completed by the Russians

1907 Picasso paints *Demoiselles d'Avignon*
1908 The Congo Free State becomes the Belgian Congo
1908 Georges Sorel writes *Reflections on Violence*

1910

1910 The Union of South Africa is formed
1911 Italy starts its conquest of Tripoli

1910 Stravinsky's *Firebird* is performed
1911 Richard Strauss writes *Der Rosenkavalier*

1913 D. H. Lawrence writes *Sons and Lovers*
1913 Proust produces Volume I of *A la Recherche du temps perdu*
1914 Matisse paints *Les Poissons rouges*

BIBLIOGRAPHY

These books were selected during the preparation of this volume for their interest and authority, and for their usefulness to readers seeking additional information on specific points.

An asterisk (*) marks works available in both hard-cover and paperback editions; a dagger (†) indicates availability only in paperback.

ART AND LITERATURE

Braun, Sidney D., ed., *Dictionary of French Literature.* Philosophical Library, 1958.
Brereton, Geoffrey, *An Introduction to the French Poets: Villon to the Present Day.* Essential Books, Fair Lawn, N.J., 1957.
Cazamian, L., *A History of French Literature.* Oxford at the Clarendon Press, 1955.
Chiari, Joseph, *Realism and Imagination.* Barrie and Rockliff, London, 1960.
Gaunt, William, *Everyman's Dictionary of Pictorial Art.* E. P. Dutton, 1962.
Giedion, Siegfried, *Space, Time and Architecture.* Harvard University Press (4th ed.), 1962.
Gloag, John, *Victorian Taste, Some Social Aspects of Architecture and Industrial Design, from 1820-1900.* A. & C. Black, London, 1962.
Gilot, Françoise, *Life With Picasso.* McGraw-Hill, 1964.
Janson, H. W., *History of Art.* Harry N. Abrams, 1962.
Rewald, John, *The History of Impressionism.* The Museum of Modern Art, 1961.
Rewald, John, *Post-Impressionism from Van Gogh to Gauguin.* The Museum of Modern Art, 1962.
Saintsbury, George, *A Short History of French Literature.* Oxford at the Clarendon Press (7th ed.), 1945.
Wilenski, R. H., *Modern French Painters,* Vol. I *(1863-1903)*; Vol. II *(1904-1938).* Vintage, 1960.
Wilson, Angus, *Emile Zola, An Introductory Study of His Novels.* William Morrow, 1952.
Wilson, Edmund, *Axel's Castle.* Charles Scribner's Sons, 1950.

GENERAL HISTORY

*Binkley, Robert C., *Realism and Nationalism 1852-1871.* Harper & Row, 1963.
*Briggs, Asa, *Victorian People: A Reassessment of Persons and Themes 1851-67.* Harper Colophon Books, 1963.
*Brinton, Crane, *The Shaping of Modern Thought.* Prentice-Hall, 1963.
*Brogan, D. W., *The French Nation.* Harper & Row, 1957.
Bruun, Geoffrey, *Nineteenth Century European Civilization: 1815-1914.* Oxford University Press, 1960.
Buckley, Jerome Hamilton, *The Victorian Temper.* Harvard University Press, 1951.
*Bury, J. B., *The Idea of Progress.* Dover, 1955.
Clough, S. B., Otto Pflanze and Stanley G. Payne, *A History of the Western World: Modern Times.* D. C. Heath, 1964.
*Cobban, Alfred, *History of Modern France,* 2 vols. Penguin, 1961.
Derry, John W., *A Short History of Nineteenth Century England: 1793-1868.* Mentor Books, New American Library, 1963.
Florinsky, Michael T., *Russia,* Vol. II. Macmillan, 1953.
*Hayes, Carlton J. H., *A Generation of Materialism 1871-1900.* Harper & Row, 1963.
Hibbert, Christopher, *Garibaldi and His Enemies.* Little, Brown, 1966.
Horne, Alistair, *The Fall of Paris.* St. Martin's, 1965.
Kirwan, Daniel J., *Palace and Hovel.* A. Allan, ed. Abelard-Schuman, London and N.Y., 1963.
Knapton, E. J., and T. K. Derry, *Europe 1815-1914.* Charles Scribner's Sons, 1956.
*Kochan, Lionel, *The Making of Modern Russia.* Penguin, 1963.
Kranzberg, Melvin, *The Siege of Paris 1870-71.* Cornell University Press, 1950.
Marder, Arthur, *The Anatomy of British Sea Power.* Shoe String, 1940.
Marlowe, John, *World Ditch: The Making of the Suez Canal.* Macmillan, 1964.
The New Cambridge Modern History. Vol. X, *The Zenith of European Power 1830-1870,* John Bury, ed. Cambridge University Press, 1964; Vol. XI, *Material Progress & World-Wide Problems 1870-1898,* F. H. Hinsley, ed. Cambridge University Press, 1962; Vol. XII, *The Era of Violence 1898-1945,* David Thomson, ed. Cambridge University Press, 1964.
Palmer, R. R., and Joel Colton, *A History of the Modern World.* Alfred A. Knopf (3rd ed.), 1965.

*Passant, E. J., *A Short History of Germany 1815-1945.* Cambridge University Press, 1962.
Pinson, Koppel S., *Modern Germany.* Macmillan, 1954.
Steinberg, Jonathan, *The Other Deterrent.* MacDonald, 1965.
Taylor, A.J.P., *The Struggle for Mastery in Europe 1848-1918.* Oxford University Press, 1965.
*Thomson, David, *England in the Nineteenth Century.* Penguin Books, 1966.
Tuchman, Barbara W., *The Proud Tower.* Macmillan, 1965.
*Wolf, John B., *France: 1814-1919.* Harper & Row, 1963.
Wood, Violet, *Victoriana, A Collector's Guide.* G. Bell & Sons, London, 1960.
Wright, Gordon, *France in Modern Times.* Rand McNally & Co., 1960.
*Young, G. M., *Victorian England: Portrait of an Age.* Oxford University Press, 1964.

SCIENCE AND TECHNOLOGY

Beatty, Charles, *De Lesseps of Suez.* Harper & Row, 1956.
Bernard, Claude, *An Introduction to the Study of Experimental Medicine.* Dover, 1957.
Derry, T. K., and Trevor Williams, *A Short History of Technology.* Oxford University Press, 1961.
Dugan, James, *The Great Iron Ship.* Harper & Brothers, New York, 1953.
Fay, C. R., *Palace of Industry, 1851.* Cambridge University Press, 1951.
*Finch, James K., *The Story of Engineering.* Doubleday, 1960.
A History of Technology, Vol. 5. Charles Singer, E. J. Holmyard, A. R. Hall, and Trevor Williams, eds. Oxford University Press, 1958.
Hobhouse, Christopher, *1851 and the Crystal Palace.* E. P. Dutton, 1950.
Luckhurst, Kenneth W., *The Story of Exhibitions.* The Studio Publications, 1951.
†Mason, Stephen, *A History of the Sciences.* Collier, 1962.
Rosen, George, *A History of Public Health.* MD Publications, 1958.
*Usher, A. P., *A History of Mechanical Inventions.* Harvard University Press, 1954. Beacon Press, Boston, 1959.
Victoria & Albert Museum, *The Great Exhibition of 1851: A Commemorative Album.* Compiled by C. H. Gibbs-Smith. Her Majesty's Stationery Office, 1964.

POLITICS

*Arendt, Hannah, *On Revolution.* Viking, 1965.
*Carr, E. H., *Michael Bakunin.* Vintage, 1961.
*Longford, Elizabeth, *Queen Victoria: Born to Succeed.* Harper & Row, 1965.
Moon, Parker T., *Imperialism and World Politics.* Macmillan, 1926.
*Postgate, Raymond W., ed., *Revolution from 1789 to 1906.* Harper & Row, 1962.
*Wolfe, Bertram D., *Three Who Made a Revolution.* Dell, 1961.

SOCIOLOGY AND ECONOMICS

*Burn, W. L., *The Age of Equipoise.* W. W. Norton, 1965.
The Cambridge Economic History of Europe, Vol. VI, *The Industrial Revolutions and After.* H. J. Habakkuk and M. Postan, eds., Cambridge University Press, 1965.
Clough, S. B., and C. W. Cole, *Economic History of Europe.* D. C. Heath, Boston, 1946.
Cole, G.D.H., *A Short History of the British Working Class Movement 1789-1947.* Hillary, 1951.
Dutton, Ralph, *The Victorian Home.* B. T. Batsford, London, 1954.
Gregg, Pauline, *A Social and Economic History of Britain: 1760-1965.* George G. Harrap (5th ed.), London, 1965.
*Hook, Sidney, *Marx and the Marxists.* D. Van Nostrand, 1955.
Lewis, John, *The Life & Teaching of Karl Marx.* Lawrence & Wishart, London, 1965.
*Morton, Frederic, *The Rothschilds.* Atheneum, 1962.
Reader, W. J., *Life in Victorian England.* Putnam, 1965.

ART INFORMATION AND PICTURE CREDITS

The sources for the illustrations in this book are set forth below. Descriptive notes on the works of art are included. Credits for pictures positioned from left to right are separated by semicolons, from top to bottom by dashes. Photographers' names which follow a descriptive note appear in parentheses. Circa is abbreviated "ca."

Cover—Photo J. H. Lartigue from Rapho Guillumette—courtesy Gernsheim Collection, University of Texas—Roger-Viollet, Paris; courtesy Fondazione Primoli, Rome.

CHAPTER 1: 8—Color lithograph from *Dickinson's Comprehensive Pictures of the Great Exhibition of 1851 from the originals painted . . . for H.R.H. Prince Albert* by Dickinson Brothers, London, 1853-1854, Cooper Union Museum Library, New York (Herbert Orth). 11—Joseph Paxton's original blotting-paper sketch for the Crystal Palace made at Derby, June 11, 1850, Victoria and Albert Museum, London. 15—Prince Albert's season ticket to the Great Exhibition of 1851, Victoria and Albert Museum, London. 17-24—Color lithograph from *Dickinson's Comprehensive Pictures of the Great Exhibition of 1851 from the originals painted . . . for H.R.H. Prince Albert* by Dickinson Brothers, London, 1853-1854, New York Public Library (Frank Lerner). 25—Color lithograph from *Industrial Arts of the Nineteenth Century at the Great Exhibition, 1851* by M. Digby Wyatt, London, 1851-1853, New York Public Library (Albert Fenn)—color lithograph from *Industrial Arts of the*

Great Exhibition of 1851 by M. Digby Wyatt, London, 1851-1853, The Mansell Collection, London; color lithograph from *Industrial Arts of the Nineteenth Century at the Great Exhibition of 1851* by M. Digby Wyatt, 1851-1853, New York Public Library (Albert Fenn). 26-27—Color lithograph from *Dickinson's Comprehensive Pictures of the Great Exhibition of 1851 from the originals painted . . . for H.R.H. Prince Albert* by Dickinson Brothers, London, 1853-1854, New York Public Library (Frank Lerner); color lithograph from *Dickinson's Comprehensive Pictures of the Great Exhibition of 1851 from the originals painted . . . for H.R.H. Prince Albert* by Dickinson Brothers, London, 1853-1854, New York Public Library (Frank Lerner)—oil color print from unbound collection *The Gems of the Great Exhibition* by George Baxter, London, 1851, Library of Congress (Herbert Orth).

CHAPTER 2: 28—Collection Tim Gidal. 33—Reprinted by permission of the Princeton University Press, copyright 1943, by Princeton University Press. 34—Culver Pictures. 36—Caricature from *The London Sketch Book,* May 1874, Radio Times Hulton Picture Library, London. 39—Eiffel

Tower under construction, 1888 (Photos Chevojon). 40-47—Drawings by Otto Van Eersel. 40-41—Top left, Bessemer converter, drawing adapted from *An Autobiography* by Sir Henry Bessemer, London, offices of "Engineering," 1905—center, Firth of Forth Bridge, drawing adapted from *Das Jahrhundert des Eisenbahn* by Wulf Schadendorf, Prestel Verlag, Munich, 1965; bottom right, the Pont Adolphe, drawing adapted from *A History of Technology* by Charles Singer, E. J. Holmyard, A. R. Hall and Trevor I. Williams (editors), Oxford University Press, Oxford, 1958. 42-43—Top right, Kelvin mirror galvanometer, drawing adapted from *Fifty Years of Electricity* by J. A. Fleming, Iliffe and Sons Limited, Dorset House, London, 1921. 44—Bottom, dynamo, drawing adapted from *A First Electrical Book for Boys*, page 164, by Alfred Morgan. Copyright 1935, 1951 Charles Scribner's Sons; renewal copyright © 1963 Alfred Morgan. Reproduced by Permission of Charles Scribner's Sons. 46-47—Right, Benz 1885 automobile, drawing adapted from original patent reproduced in *75 Jahre Motorisierung des Verkehrs*, Stuttgart, 1961.

CHAPTER 3: 48—Bank notes, 1850-1914, courtesy Chase Manhattan Bank Money Museum, New York (Donald Miller). 50—Archiv für Kunst und Geschichte, Berlin. 55—Brown Brothers, New York. 59—*Main Hall of the Old Stuttgart Railroad Station* by Hermann Pleuer, oil on canvas, 1905, Staatsgalerie, Stuttgart (Walter Sanders). 60-61—*The Railroad Bridge at Argenteuil* by Claude Monet, oil on canvas, ca. 1873, Musée du Louvre, Paris (Giraudon, Paris); *Saint-Lazare Station* by Claude Monet, oil on canvas, 1877, Fogg Museum, Harvard University, Wertheim Collection (Scala, Florence). 62—*Boulevard des Capucines* by Claude Monet, oil on canvas, 1872, courtesy Mrs. Marshall Field, New York (Propriété des Editions d'Art Albert Skira, Geneva). 63—*The Print Collector* by Honoré Daumier, oil on canvas, ca. 1857, Musée du Petit Palais, Paris (Bulloz, Paris)—*The Milliner's Shop* by Edgar Degas, oil on canvas, ca. 1882, courtesy The Art Institute, Chicago (Arthur Siegel). 64—*Café Concert at Les Ambassadeurs* by Edgar Degas, oil on canvas, ca. 1876, Musée des Beaux Arts, Lyons (Giraudon, Paris). 65—*At the Moulin Rouge* by Pierre Bonnard, oil on canvas, 1896, courtesy Wright Ludington Collection, Santa Barbara (Propriété des Editions d'Art Albert Skira, Geneva). 66-67—*Train in the Country* by Claude Monet, oil on canvas, ca. 1870, Musée du Louvre, Paris (Giraudon, Paris)—*At the Races* by Edgar Degas, oil on canvas, 1862, Musée du Louvre, Paris (Istituto Italiano d'Arti Grafiche, Bergamo); *La Grenouillère* by Claude Monet, oil on canvas, 1869, The Metropolitan Museum of Art, New York, bequest of Mrs. H. O. Havemeyer, 1929, the Havemeyer Collection (Frank Lerner). 68—*Dieppe* by Camille Pissarro, oil on canvas, 1892, courtesy Alfred Schwabacher, New York (Frank Lerner). 69—*Hôtel des Roches Noires, Trouville* by Claude Monet, oil on canvas, 1870, courtesy Laroche Collection, Paris (Propriété des Editions d'Art Albert Skira, Geneva). 70-71—*The Beach at Sainte Adresse* by Claude Monet, oil on canvas, date unknown, The Metropolitan Museum of Art, New York, bequest of William Church Osborn, 1951 (Frank Lerner).

CHAPTER 4: 72—Photograph by John Thomson from *Street Life in London* by John Thomson and Adolphe Smith, London, 1877, courtesy George Eastman House, Rochester. 76—Housing plans by Prince Albert, President of the Society for Improving the Conditions of the Labouring Classes, 1851 (Ronan Picture Library, Newmarket). 81—Membership card, International Working Men's Association, 1864 (Bettmann Archive). 83—*Outbreak* by Käthe Kollwitz, charcoal drawing, 1903, Deutsche Akademie der Kunst, Berlin (Deutsche Fotothek, Dresden). 84-85—*Plowman and Wife* by Käthe Kollwitz, lithograph, 1902, courtesy Galerie St. Etienne, New York (Frank Lerner). 86-87—*Home Work* by Käthe Kollwitz, drawing, 1909, Kunsthalle, Bremen (Hermann Stickelmann); *Midday Meal* by Käthe Kollwitz, tempera and charcoal, 1909, courtesy Mr. and Mrs. William Lincer, New York, copyright Galerie St. Etienne, New York (Frank Lerner). 88-89—*Hamburg Tavern* by Käthe Kollwitz, soft-ground etching, 1901, courtesy Leonard Spitalnik, New York (Frank Lerner); *Mother with Child in her Arms* by Käthe Kollwitz, etching, 1919, courtesy Galerie St. Etienne, New York (Frank Lerner). 90-91—*Need* by Käthe Kollwitz, lithograph, 1907, courtesy Galerie St. Etienne, New York (Frank Lerner); *Pietà* by Käthe Kollwitz, lithograph, 1903, Staatliche Museen zu Berlin, Photographische Abteilung. 92-93—*Conspiracy* by Käthe Kollwitz, lithograph, 1898, courtesy Galerie St. Etienne, New York (Frank Lerner); *Marching Weavers* by Käthe Kollwitz, etching, 1897, courtesy Galerie St. Etienne, New York (Frank Lerner)—*Storming the Gates* by Käthe Kollwitz, etching, 1897, courtesy Galerie St. Etienne, New York (Frank Lerner). 94-95—*Uprising* by Käthe Kollwitz, etching, 1899, courtesy Galerie St. Etienne (Frank Lerner).

CHAPTER 5: 96—*Otto von Bismarck* by Franz von Lenbach, oil on canvas, 1895, Bismarck-museum, Friedrichsruh, Germany (Friedrich Rauch). 107—*Piccoli Garibaldini* by G. Toma, oil on canvas, 1862, Museo del Risorgimento, Bergamo (David Lees). 108—*Garibaldi and Major Leggero Carrying Anita through the Marshes* by P. Bouvier, oil on canvas, ca. 1890, Museo del Risorgimento, Milan (David Lees). 109—*Garibaldi in 1845* by J. Malinski, oil on canvas, Museo del Risorgimento, Milan (David Lees)—Garibaldi's Colt revolver, Museo del Risorgimento, Turin (David Lees); *A Bivouac* by G. Induno, oil on canvas, ca. 1859, Museo del Risorgimento, Milan (David Lees). 110-111—*Embarkation of the Thousand at Quarto* by G. Induno, oil on canvas, ca. 1860, Museo del Risorgimento, Milan (David Lees); *Debarkation of Garibaldi at Marsala* by G. Induno, oil on canvas, ca. 1860, Museo del Risorgimento, Turin (David Lees)—Garibaldi's sword, courtesy David Cugini, Bergamo (David Lees). 112-113—*Battle of Calatafimi* by R. Legat, oil on canvas, date unknown, Museo del Risorgimento, Milan (David Lees); *Garibaldi at the Defense of Rome* by unknown artist, oil on canvas, ca. 1851, Museo del Risorgimento, Milan (David Lees)—Garibaldi's saddle, Museo del Risorgimento, Turin (David Lees). 114-115—*King and Queen of Naples under Siege at Gaeta* by C. Bossoli, oil on canvas, ca. 1861, Museo del Risorgimento, Turin (David Lees)—*The Tuscan Delegation Presents the Act of Annexation to Victor Emmanuel II* by G. Mochi, oil on canvas, ca. 1861, Florence Museum, Florence (David Lees); *Giuseppe Garibaldi* by D. Induno, oil on canvas, ca. 1850, Museo del Risorgimento, Milan (David Lees). 116-117—Garibaldi's cap, Museo del Risorgimento, Turin (David Lees); *Meeting at Teano of Garibaldi and Victor Emmanuel II* by Sebastiano de Albertis, oil on canvas, ca. 1886, courtesy Marchese Luigi Medici del Vascello, La Mandria, Turin (David Lees).

CHAPTER 6: 118—Culver Pictures. 123—Hoover Institute on War, Revolution, and Peace, Stanford, California (Jon Brenneis). 127—Bibliothèque Nationale, Paris. 128—Archives Editions Robert Laffont, Paris—Bibliothèque Nationale, Paris. 129—Archives Editions Robert Laffont, Paris—Illustrated London News—Archives Editions Robert Laffont, Paris. 130—Archives Editions Robert Laffont, Paris—Illustrated London News; The Mansell Collection, London; Archives Editions Robert Laffont, Paris; Archives Editions Robert Laffont, Paris—Archives Editions Robert Laffont, Paris. 131—Archives Editions Robert Laffont, Paris—Archives Editions Robert Laffont, Paris; Archives Editions Robert Laffont, Paris—Culver Pictures Incorporated. 132—Bibliothèque Nationale, Paris—Bibliothèque Nationale, Paris; Archives Editions Robert Laffont, Paris—Archives Editions Robert Laffont, Paris. 133—André-Molinier, Paris—Radio Times Hulton Picture Library, London—Radio Times Hulton Picture Library, London. 134—Archiv für Kunst und Geschichte, Berlin; Archives Editions Robert Laffont, Paris—Archives Editions Robert Laffont, Paris; The Bettmann Archive. 135—Radio Times Hulton Picture Library, London—Radio Times Hulton Picture Library, London—Radio Times Hulton Picture Library, London. 136—Radio Times Hulton Picture Library, London—Library of Congress (Frank Lerner)—Radio Times Hulton Picture Library, London; Radio Times Hulton Picture Library, London. 137—Radio Times Hulton Picture Library, London—Giraudon, Paris.

CHAPTER 7: 138—*Claude Monet Painting* by John Singer Sargent, oil on canvas, ca. 1889, courtesy the Trustees of the Tate Gallery, London, gift of Emily Sargent and Mrs. Ormond. 141—*Battle of the Schools* by Honoré Daumier, caricature from *Le Charivari*, 1855, from *The History of Impressionism* by John Rewald, The Museum of Modern Art, New York, 1961 (Charles Phillips). 149—Lamp by Raoul Larche, courtesy Joseph H. Heil, New York, and wallpaper by William Morris, courtesy W.H.S. Lloyd Co., Inc., New York (Arnold Newman). 150-151—Furniture and art objects courtesy Joseph H. Heil, Lillian Nassau Antiques, James Benton and William Leazer, New York, and wallpaper by William Morris courtesy W.H.S. Lloyd Co., Inc., New York (Arnold Newman). 152-153—Wallpaper by William Morris and art objects courtesy Joseph H. Heil, New York (Arnold Newman). 154-155—Brooch by René Lalique, courtesy Gulbenkian Foundation, Lisbon (Sabine Weiss)—vase by Louis Comfort Tiffany and stickpins courtesy Joseph H. Heil, New York (Arnold Newman); jewelry and small objects courtesy James Benton and William Leazer, New York (Arnold Newman). 156-157—Books courtesy Joseph H. Heil, James Benton and William Leazer, Lucien Goldschmidt and J. N. Bartfield Fine Book Studios, New York (Arnold Newman); type designs by Otto Eckmann, Museum für Kunst und Gewerbe, Hamburg (Sandak). 158—Poster by Jacques Villon, The Museum of Modern Art, New York (Sandak); poster by Alfons Mucha, Victoria and Albert Museum, London (Heinz Zinram). 159—Poster by Eugène Verneau, Courtesy Library of the Musée des Arts Décoratifs, Paris (Heinz Zinram)—poster by Kolomon Moser, The Museum of Modern Art, New York (Sandak); poster by Alfons Mucha, Griffin Gallery, New York (Heinz Zinram). 160—*Horses of Neptune* by Walter Crane, oil on canvas, 1892, Bayerische Staatsgemäldesammlung, Munich (Joachim Blauel)—*Moonlight* by Edvard Munch, woodcut, 1896, Graphische Sammlung, Munich (Joachim Blauel). 161—*Mrs. Flöge* by Gustav Klimt, oil on canvas, 1902, Historisches Museum der Stadt Wien (Erich Lessing from Magnum). 162-163—Roof of the Milá house in Barcelona by Antonio Gaudí (N. R. Farbman).

CHAPTER 8: 164—*Ermenonville* by Henri Gervex, oil on canvas, 1909, courtesy Collection Comte de Hamal, Paris (Pierre Boulat). 166—Radio Times Hulton Picture Library, London. 169—Wiener Männergesang-Verein. 173—Courtesy Dr. Jacques Emile-Zola Collection, Paris. 174—Historical Pictures Service, Chicago; courtesy Harvard Theatre Collection; photograph by Bert from *Le Ballet* by Boris Kochno, Librairie Hachette, Paris, 1954 (Alan Clifton)—Library of Congress (Frank Lerner); Historisches Bildarchiv, Bad Berneck. 175—Historical Pictures Service, Chicago—Library of Congress (Frank Lerner); Sovfoto-Historical Pictures Service, Chicago. 176-177—Historical Pictures Service, Chicago; Historisches Bildarchiv, Bad Berneck; The Mansell Collection, London—Library of Congress (Frank Lerner); Library of Congress (Frank Lerner); Archiv für Kunst und Geschichte, Berlin; courtesy George Eastman House, Rochester—The Mansell Collection, London; Radio Times Hulton Picture Library, London; Historical Pictures Service, Chicago. 178—Library of Congress (Frank Lerner)—Henri Laurens, Paris; Archiv für Kunst und Geschichte, Berlin—René Char—courtesy Ludwig Charell. 179—G. Cres et Cie.; Bonney, Paris—Bildarchiv, Österreichische Nationalbibliothek, Vienna; Collection J. Mucha, Prague; Pictures Inc. 180-181—Paul Nadar; The Bettmann Archive; Culver Pictures; Historisches Bildarchiv, Bad Berneck; Historisches Bildarchiv, Bad Berneck—Historisches Bildarchiv, Bad Berneck; Library of Congress (Frank Lerner); Library of Congress (Frank Lerner); Library of Congress (Frank Lerner); Historisches Bildarchiv, Bad Berneck. 182—Historical Pictures Service, Chicago—Historisches Bildarchiv, Bad Berneck; Historical Pictures Service, Chicago. 183—Wide World Photos Inc.; Pictures Inc.—Library of Congress (Frank Lerner); The Mansell Collection, London.

ACKNOWLEDGMENTS

For the help given in the preparation of this book, the editors are particularly indebted to Professor Andrew Whiteside, Department of History, Queens College, New York. The editors also express their gratitude to the following advisors: Frederick Kilgour, Yale University Library; Dr. Otto Kallir, Galerie St. Etienne, New York City; Professor Daniel A. Greenberg; Dr. Allen McConnell; Milton Kaplan and Dr. Allan Fern, Prints and Photographs Division, Library of Congress; Professor Theodor Reff, Department of Art History and Archeology, Columbia University; Joseph H. Heil; James Benton; W.H.S. Lloyd Co., Inc.; Lucien Goldschmidt; Leopoldo Marchetti, Director, Museo del Risorgimento, Milan; Raccolta Civica de Stampe Bertarelli, Milan; Piero Pieri, Director, and Teresa Conte, Museo Nazionale del Risorgimento, Turin; Marchese Luigi Medici del Vascello, La Mandria, Turin; Alberto Ghisalberti, Director, Emilia Morelli and Mariuccia Belloni-Zecchinelli, Istituto del Risorgimento, Rome; Carlo Pietrangeli, Director, and Lucilla Negro, Fondazione Primoli, Rome; Gino Doria, Naples; Annamaria Bonucci, Director, Museo di San Martino, Naples; Bianca Montale, Istituto Mazziniano, Genoa; Pietro Aranguren, Director, Istituto del Risorgimento, Florence; André Jammes; Max Terrier, Conservateur, Musée National de la Voiture et du Tourisme, Compiègne; M. D. Lengellé, La Vie du Rail, Paris; Adeline Cacan, Conservateur, Petit Palais, Paris; Staatliche Museen zu Berlin, Kupferstichkabinett; Dr. Hans Kollwitz, Berlin; Bayerische Staatsgemaeldesammlung, Munich; Historisches Bildarchiv, Bad Berneck; Deutsche Fotothek, Dresden; Archiv für Kunst und Geschichte, Berlin; Direktor Franz Glück, Historisches Museum der Stadt Wien; Hofrat Doktor Hans Pauer, Bildarchiv, Österreichische Nationalbibliothek, Vienna; Charles Gibbs-Smith, Victoria and Albert Museum, London.

INDEX

This symbol in front of a page number indicates a photograph or painting of the subject mentioned.

189

Perkin, William Henry, 31-32
Persia, revolution of 1909, 125
Petit Journal, Le (Paris), 82
Petroleum industry, 124
Philippine Islands, *map* 105
Philosophy, 29; Nietzsche, 166-167
Photoengraving, 38
Photography, 14, 39, *173-183
Physics, 30
Picasso, Pablo, 145, 161
Picture of Dorian Gray, The, Wilde, 157
Piedmont, 99, *map* 100, 101. *See also* Sardinia-Piedmont, Kingdom of
Pissarro, Camille, 142, 145; painting by, *68
Pittsburgh, Pa., 41
Pius IX, Pope, 16
Plastics, 32, 39
Plehve, Wenzel von, 122
Poetry, Symbolist, 145-146, 148
Pogroms, anti-Semitic, 75, 122
Pointillism, 147
Poland: nationalist movement, 98; persecution and emigration of Jews, 75
Poles, in Austrian Empire, 167
Political parties, 77; socialist, 80, 81, 121, 125, 126
Pont Adolphe, Luxembourg, *41
Pontecorvo, Italy, *map* 100
Popular sovereignty, 98, 102, 105. *See also* Representative government
Population growth, European Continent, 49-50, 74
Porcelain, 15, 24, *25
Port Arthur, Japanese seizure of, 123
Port Sunlight, England, 77-78
Portugal, colonial empire of, *map* 104-105
Portuguese Guinea, *map* 104
Post-Impressionism, 146-148; defined, 145
Poster art, *158-159
Potato famines, Ireland, 55, 75, 126
Potemkin mutiny, 125
Power, sources of, 10, 12, 13-14, 34, 35, *44-45
Power loom, 13, 14
Powers, Hiram, *The Greek Slave,* *26-27, 139
Prefabrication, 11
Printing presses, 13, 38
Proletariat, 56, 80. *See also* Working class
Protectionism, 57
Proudhon, Pierre-Joseph, 121
Prussia, Kingdom of, 102, *map* 103, 104-105; Bismarck's war policy, 102-103, 128; militarism, 102, 105; Seven Weeks' War with Austria (1866), 102, *map* 103; war with Denmark (1864), 102. *See also* Franco-Prussian War
Psychology, 29; Freud, 166, 168
Public health, 36-37, 75, 77, 82, 90
Public libraries, 38
Public welfare. *See* Social reform
Punch, magazine, 11

R

Rabies vaccination, *34
Racial superiority, notions of, 106
Radio: Marconi's invention, 38; Maxwell's postulation of waves, 32
Railroads, *50, 51, *60, *68; Berlin-Baghdad, 166; financing, 53, 54; stations, *59-61; suburban, 66-67; Trans-Siberian, 124; use in warfare, 102, 103; West European countries, 12, 13-14, 15, 50, 51, 54, 60
Raw materials, 13, 38, 51, 56
Rayon, 32
Realism: defined, 145; literature, 143, 145; painting, 140-141, 143, 179
Realpolitik of Bismarck, 79, 101
Reaper, mechanical, 15, 30
Recreation, *63-71; municipal efforts, 82; outdoor, *62, *66-67, *70-71; Paris entertainment, *64-65; resorts, *68-71
Red Flag Act, Great Britain, 47
Reform movements, 50, 77-78
Religion: effect of industrialization on, 16; effect of science on, 29, 37
Renaissance painting, 142

Renoir, Auguste, 140, 142, *179
Representative government, 50, 98, 99, 104-105, 125, 129, 167
Rerum novarum, encyclical, 80-81
Resorts, *68-71
Revolution(s): China (1911), 125; of 1830, 98, 99; of 1848, 98, 99, 102, 108; of 1860 in Italy, 100-101; Marxist-socialist call for, 79-80, 81, 120, 121-122; nationalist call for, 98; Paris Commune, 73, 119-120, 121, 123, *127, *133-137; Persia (1909), 125; Russia (1905), 73, *118, 123-125; Russia (1917), 73, 182; Turkey (1908), 125
Révue Wagnerienne, magazine, 146
Rhodesia, *map* 104
Río de Oro, *map* 104
Río Muni, *map* 104
Riso, Francesco, 100
Rodin, Auguste, *178
Roentgen, Wilhelm Konrad, 39
Romagna, in united Italy, *map* 100, 115
Roman Catholic Church, under Leo XIII, 80-81
Romania, liberation from Turkish rule, 101
Romanians, in Austrian Empire, 167
Romanov dynasty, 172
Romanticism, 140-141
Rome: annexation by Italy, *map* 100, 101; Papal States, 99, *map* 100, 115
Roosevelt, Theodore, 167
Rothschild family, 54-56
Rouault, Georges, 145
Rougon-Macquart, Les, Zola, 144
Rousseau, Henri, *178
Rousseau, Jean Jacques, 97
Royal Academy of Art, London, 140
Royalty, European, *182-183
Rubber, 15, 51, 56
Rudolf, Archduke of Austria, *183
Ruhr industry, 12
Ruskin, John, quoted, 16
Russia: *map* 103; abolition of serfdom, 74, 121, 124, 144; and Balkan crises, 170-171; Duma (parliament), 125; and events leading to World War I, 172; first Soviet, 125; and Franco-Prussian War, 102, *cartoon* 128; grain exports, 57; industrialization, 124; Jewish pogroms, 122; lack of social reform, 81, 94, 121, 122, 124; literature, 144-145; peasant uprisings, 84, 125; protectionism, 57; railroads, 124; Revolution of 1905, 73, *118, 123-125; Revolution of 1917, 73, 182; in Russo-Japanese War, 123, 125; Social Revolutionaries, 122; strikes, 124-125; in Triple Entente, 168; urban migration, 74
Russo-Japanese War, 123, 125
Ruthenians, in Austrian Empire, 167

S

St. Petersburg: first Soviet, 125; machine industry, 124; Revolution of 1905 in, 123-124, 125
Saint-Simon, Comte de, 79
Salon des Beaux-Arts, Paris, 140, 141-142
Salon des Refusés exhibit, 142
Salt, Sir Titus, 77
Saltaire, England, 77
Sand, George, *180
Sanitarians, 36
Sanitation, 36, 75, 82
San Marino, *map* 100
Sarajevo assassination, 171-172
Sarawak, *map* 105
Sardinia-Piedmont, Kingdom of, 99, *map* 100, 101
Sargent, John Singer, painting by, *138
Savery, Thomas, 12
Savoy, *map* 100
Savoy, House of, 99; emblem, *113
Saxony, Kingdom of, *map* 103
Schleswig, Duchy of, 102, *map* 103
Science, 10, 29-38, 166; applied, 29-36; basic tenets of modern, 29; exhibits, *33; laboratories, 29, 35, 50; team work, 29
Scott, Sir Walter, 144
Sculpture, 179; Crystal Palace exhibit, *26-27, 139

Seaside resorts, 59, *68-71
Second International, 81
Sedan, battle of, *map* 103, 104, 119, 128
Serbia, 168; and assassination at Sarajevo, 171-172; in Balkan crisis (1908), 170; in Balkan Wars (1912, 1913), 170, 171
Serbs, in Austrian Empire, 167
Serfdom, abolition of, 73-74, 121, 124, 144
Sergius Alexandrovitch Romanov, Grand Duke, assassination of, 122
Seurat, Georges, 145, 147, 148, 161
Seven Weeks' War between Austria and Prussia (1866), 102, *map* 103
Severini, Gino, 145
Sèvres porcelain, *25
Sewing machine, invention of, 30
Shaw, George Bernard, 80
Sheffield, England, 41
Shelley, Percy Bysshe, 144
Shipping: British supremacy, 12, 168-169; canals, 52-53; German ascendancy, 168-169; introduction of iron hulls, 13; introduction of steam, 13, 52
Sicily, *map* 100; 1860 uprising in, 97, 100; Garibaldi's campaign in, 97, 100, 110, *111-113, 115. *See also* Two Sicilies, Kingdom of the
Siemens, Werner, 15
Sierra Leone, *map* 104
Singapore, *map* 105
Sisley, Alfred, 142, 145
Slave trade, abolition of, 57
Slavic nationalism, 106, 167-168, 170, 171-172
Slovaks, in Austrian Empire, 167
Slovenes, in Austrian Empire, 167
Slums: clearance, 82; poverty, *72, 73, 75, 90
Social Contract, The, Rousseau, 97
Social Darwinism, 105
Social Democratic party, Austria, 121
Social Democratic party, Germany, 80, 126
Social Democratic party, Russia, 125
Social reform and legislation, 81-82, 94, 124; Austria, 167; demands for, 58, 77-78, 80-81, 119, 120-125; Germany, 79, 80; Great Britain, 77-78; reform-minded employers, 51, 77-78
Social Revolutionary party, Russia, 122
Social structure: before Industrial Revolution, 73, 80; class gulf, 56, 73, 75-76, 81-82, 120-121, 122, 126; English, compared with Continental European, 12-13; moneyed aristocracy, 53-56, 80; rise of middle class, 12-13, 56, 80; upper classes, 12, 56, 73, 75-76, 80; working class, 56, 73-74, 75-77, 80, 81-82, *83-95, 120-121
Socialism, 78-80, 81; Christian, 81; in Germany, 79, 80, 126
Socialist Labor party, France, 80
Sociology, 29
Solferino, Italy, *map* 100; battle of, 99
Solomon Islands, *map* 105
Somaliland, Italian, British, French, *map* 104
South Africa, *map* 104
South America: first cable to, 43; railroads, 51; trade, 51
South West Africa, *map* 104
Soviet of workers, first, 125
Spain: agrarian, 50; colonial empire of, *map* 104-105; Crystal Palace exhibit of, 15, 24, *25; and Franco-Prussian War, *cartoon* 128; nationalists, 98; protectionism, 57; railroads, 51; succession dispute of 1870, 102-103
Spices trade, 51
Spinning jenny, 14
Sports, *66-67
Sportsman's Sketches, A, Turgenev, 144
Stanley, H. M., 56
Steam engine, 13-14, 30, 53; invention of, 12, 30; use in production of electricity, 34, 35, 45
Steam locomotive, 13-14, 30
Steam turbine, 35
Steamship, 13, 30, 38, 52
Steamship lines, financing, 53-54
Steel industry: Bessemer process, *40, 41; open hearth furnace, *40; production and uses, 30, 41

Steel cannon, 15, 41, 102, 103
Stephenson, George, 14
Stock companies, 53-54
Stowe, Harriet Beecher, 144
Strauss, Johann, waltz score by, *169
Strikes, 77, 78, 83, 121, 123, 124-125
Stuttgart, Germany, railroad station, *59
Suburbs, 66-67
Suez Canal, 52-53, *55, 56
Sugar trade, 51
Sumatra, *map* 105
Supply and Demand, Natural Law of, 76
Sweden, *map* 103; socialist movement, 80
Switzerland, *map* 100, *map* 102-103; compulsory education, 32; Crystal Palace exhibit of, 15
Symbolism, 145-146, 148
Synthetics industry, 31-32

T

Tariffs, 57
Tea trade, 51
Teano, *map* 100, 106
Technology, 30, 35, 38. *See also* Engineering; Inventions
Telegraph, 15, 38, 51; invention of, 33; transoceanic, 15, 38, *42-43, 52; wireless, 39
Telephone, 39, *45
Textile industry, 12, 14, *25, 124; dyes, 31-32; labor, 74; mechanization, 14; synthetics, 32
Thackeray, William Makepeace, 143, 144
Thames River, pollution of, 90
Thermodynamics, laws of, 30-31, 38
Thiers, Adolphe, 99, *131, 133, 134, 135, 136
Thomas, Clément, *133
Thomson, Sir Joseph, 38
Thousand, The, of Garibaldi, 100, 112
Three Musketeers, The, Dumas, 175
Thuringian States, *map* 103
Tiffany, Louis Comfort, 150
Tiffany lamp, *150
Tiffany vase, *155
Timor, *map* 105
Togoland, *map* 104
Tolstoy, Leo, 144-145
Tories (Britain), and labor legislation, 77
Toulouse-Lautrec, Henri de, 150, 158, *178
Trade, international, 38, 49-51, 53; agricultural products, 50, 51, 57; British share of, 12; concept of Free, 16, 57, 76; European supremacy, 49; German ascendancy, 79, 166; industrial products, 38, 49, 51, 56; protectionism, 57; raw materials, 38, 51
Trade Union Act (1871), Great Britain, 78
Trade unions, 77, 78, 81, 93, 94, 121
Transatlantic cable, 15, 38, *42-43, 52
Transpacific cable, 43
Transportation, 10, 13, 47, 51-52. *See also* Railroads; Shipping
Trans-Siberian Railroad, 124
Travel, 10, *59-61; increase of speeds, 13, 14, 59; transatlantic, 13
Treaty of Bucharest, 171
Triple Alliance (Austria-Germany-Italy), 168
Triple Entente (Britain-France-Russia), 168, 170, 172
Trochu, Louis, *129, 130, *132
Trotsky, Leon, 125, 126
Tschaikovsky, Peter Ilyich, *174, 175
Tuberculosis, 75
Tunis, *map* 104; Crystal Palace exhibit of, *26
Turbines, 35, 45
Turgenev, Ivan, 144, 145
Turin, Italy, *map* 100, 101
Turkey. *See* Ottoman Empire
Tuscany: Grand Duchy of, 99; in united Italy, *map* 100, *114, 115
Two Sicilies, Kingdom of the, 99; in united Italy, *map* 100, 114
Typhoid, 37, 75
Typhus, 75

✕✕✕

PRODUCTION STAFF FOR TIME INCORPORATED

John L. Hallenbeck (Vice President and Director of Production),
Robert E. Foy and Caroline Ferri
Text photocomposed under the direction of Albert J. Dunn